ROUTLEDGE LIBRARY 1
GLOBAL TRANSPORT PL

Volume 13

THE RAPID TRANSIT RAILWAYS OF THE WORLD

THE RAPID TRANSIT RAILWAYS OF THE WORLD

HENRY F. HOWSON

Routledge
Taylor & Francis Group

LONDON AND NEW YORK

First published in 1971 by George Allen & Unwin Ltd

This edition first published in 2021
by Routledge
2 Park Square, Milton Park, Abingdon, Oxon OX14 4RN

and by Routledge
52 Vanderbilt Avenue, New York, NY 10017

Routledge is an imprint of the Taylor & Francis Group, an informa business

British Library Cataloguing in Publication Data
A catalogue record for this book is available from the British Library

ISBN: 978-0-367-69870-6 (Set)
ISBN: 978-1-00-316032-8 (Set) (ebk)
ISBN: 978-0-367-74108-2 (Volume 13) (hbk)
ISBN: 978-0-367-74114-3 (Volume 13) (pbk)
ISBN: 978-1-00-315611-6 (Volume 13) (ebk)

Publisher's Note
The publisher has gone to great lengths to ensure the quality of this reprint but
points out that some imperfections in the original copies may be apparent.

Disclaimer
The publisher has made every effort to trace copyright holders and would welcome
correspondence from those they have been unable to trace.

THE RAPID
TRANSIT RAILWAYS
OF THE WORLD

BY

HENRY F. HOWSON

LONDON
GEORGE ALLEN & UNWIN LTD
RUSKIN HOUSE   MUSEUM STREET

First Published in 1971

© George Allen & Unwin Ltd, 1971

ISBN 0 04 385056 1

Printed in Great Britain
in 11 point Plantin type
by Cox & Wyman Ltd
London, Fakenham & Reading

FOREWORD

During a world tour in 1964 and on subsequent journeys the author visited, in a private capacity, many of the cities mentioned in this book, and met many of their transport officials. Some, especially in Japan, have seen their particular metropolitan rapid transit system grow from a few miles of route into an extensive rail network carrying tens and in some cases hundreds of millions of passengers annually. On the other hand there have been officials who over the years have worked on rail rapid transit schemes for their cities, only to find that public backing for such projects, although stronger with each successive vote, was still not enough for a necessary majority approval to commit their communities to years of construction work, and to find the large amount of capital to pay for it.

Nevertheless, the trend is towards more urban underground railways in world cities; and whereas such a book as this written twenty years ago would have listed only about twenty operating systems, today the number of such systems in being or being built is nearer 60.

The information that follows was obtained partly by personal observation, but mostly by follow-up correspondence with city and regional authorities, to each of whom the author has written his thanks, but expresses it again here, collectively. He wishes also to reiterate to any design specialist who chances to read this book his appreciation of the inexorable march of progress, that sometimes makes a carefully designed piece of equipment obsolete before ever it reaches the production stage. In the nature of things an undertaking as big as an underground railway can only gradually assimilate the absolutely new.

Finally, he feels certain that the travelling public, busy and concerned with their own affairs, are never likely to openly enthuse about their particular Metro, but they appreciate it just the same.

CONTENTS

PART THREE: RAPID TRANSIT SYSTEMS PLANNED OR PROPOSED

CONTENTS

CONTENTS

ILLUSTRATIONS

PART ONE

Above : Vienna: tram descending under Getreidmarkt on circular route; *below :* Vienna: Schwedenplatz Station on the Stadtbahn.

2. *Above:* Paris: pneumatic train at Etoile, Metro station; *below:* Paris new rolling stock at Metro Line 3.

INTRODUCTION

The term 'Rapid Transit' has been defined as meaning 'any form of public transport separate from other transport, operating at high frequency to carry large numbers of passengers at consistently high speeds'. The term is probably of American origin and has not been in general use until recent years, but has now come into wide use to describe the underground-surface railways, individually known as the Subway, Metro, Underground or U Bahn, that are being built or planned increasingly in and around large cities.

This surge of activity seems to be based on assumptions that cities will continue to attract people, and city centres will continue to attract large day-time populations coming in from the outer suburbs. The first assumption seems at least to be reasonable, since throughout history urban life has justified itself by offering employment, education and a wider social life.

The basic reason for this immensely costly urban railway construction now under way becomes apparent only when one tries to consider what alternative means there are for moving large numbers of people into and out of cities. It may well be argued that elaborate systems of urban motorways, at elevated or sub-surface level, offer the most economic means of distribution for both passengers and merchandise within city and metropolitan areas. Using these road systems, commuters can, and do, travel independently by automobile from their homes to their offices or factories; and merchandise, broken down into suitably sized consignments, similarly and swiftly reaches its destination.

It does in fact happen like this. Major cities without urban mass transit rail systems have neither choked to death, nor has their traffic slowly ground to a halt, as was grimly foretold, simply because frustrations and economic losses resulting from congested central streets have in themselves imposed some sort of regulation and orderly movement.

B

17

But even the most modern and sophisticated of such cities are not immune from the ill effects of an overdependence on the motor vehicle. Most of them appear to have attracted or generated more surface traffic than their thoroughfares can cope with. Automobile traffic in Caracas, for instance, is reported to have semi-congealed during recent years into a continuous 7 mph flow, maintained over long periods – and there are many others similarly affected. London, Los Angeles, Tokyo, Chicago – quote any large city of like prominence, in fact, and it will concede to a traffic problem: not one arising from any deliberate indulgence towards the motor vehicle, but one that simply grew as a by-product of evolution, in this case the evolution of individual mechanised transport from the horse-drawn variety.

City traffic authorities now generally accept that something more than a system of urban highways is desirable to cope with a daily surge of commuters, shoppers and the like that may swell a day-time population to double that of a night-time population. The ideal for planners is some form of mass transport that at peripheral points will channel traffic into well defined paths, preferably condensing it from 100 vehicles carrying 200 passengers into one vehicle carrying them all – and at central points to discharge them to find their own way on foot, perhaps assisted by moving walkways, to their respective destinations. For the travelling public this media must essentially be swift, safe, reasonably comfortable and economically acceptable. In short, it must be attractive enough to wean them away from their own autos.

London Transport, in opening its new Victoria Line, thinks it has found the answer. It states 'with more passengers, particularly commuting motorists, going by Underground, the new line will bring relief to traffic-congested streets above . . . .'. In comparing the double-tracked Victoria Line with an imaginary motorway of equal carrying capacity, its designers calculated that the Tube's total capacity when completed, with trains running at two-minute intervals, will be 25,000 passengers per hour in either direction – and that a motorway with comparable total capacity would need no less than eleven traffic lanes. Similar calculations in America are even more favourable towards rapid transit, and other countries have presumably reached the same conclusions since, after weighing capital costs, economic loss from temporary disruption during construction, availability of manpower and materials and the fact that rapid transit railways tend to be under-employed during offpeak hours: all this against short and long term transportation requirements – and almost certain long term benefits – they have decided in favour of the rapid transit urban railway.

The advantages of rapid transit might be summarized as follows: Such an urban railway or system of railways of conventional design, by assuming the character of an additional artery or arteries not subject to congestion, absorbs much of the traffic normally using surface routes and thereby alleviates surface congestion. An urban railway occupies less land than a highway of comparable capacity, and less still if it is raised on aerial structure in outer areas and descends below-ground within city confines. In the long run such a railway is likely to be a social and economic attribute to a city, and to more than repay its cost in revenue plus indirect benefits.

Rapid Transit around the World

The very earliest of urban railways were transport entities in themselves, the bulk of whose passengers were pedestrian at either end of their journeys. Today, such railways work integrally with other forms of transport, principally with feeder bus services, but increasingly with 'park and ride' patrons who drive their cars or autos to outer stations and sometimes complete their journeys with a short 'in-town' bus trip. The motor truck or lorry and the automobile have in fact become most important cohesive factors in city life, the former by facilitating and speeding distribution of commodities, and the latter by providing a degree of mobility without which a city dweller's life could become hum-drum. For the foreseeable future, therefore, motor transport will probably be the prime influence in city traffic planning, in which the Metro, Subway or Underground will play its important part.

So many suggestions or proposals for new metropolitan railways in different parts of the world are coming up for consideration, or have led to detailed planning, that it is next to impossible to keep track of them all. Most of them are linked with comprehensive plans for city development, the latter in turn often stemming from great changes that are taking place in the productive and social life of whole countries, particularly the 'newer' countries. For example, in Brazil, São Paulo is fast expanding in consequence of its position at the centre of the principal industrial region of Latin America. From less developed areas of Brazil it attracts workers seeking and obtaining better living conditions through employment in commercial and industrial enterprises. Population has grown fantastically over recent years and is now more than 5½ millions in the municipal area and seven millions in the Greater São Paulo area, embracing several satellite towns.

A long sustained building boom has created a new São Paulo with innumerable high buildings, and with it a public transport problem. This, it is reported, is to be alleviated by the building of an underground railway system at an estimated cost of $380m. Across the Pacific 6,000 miles east of South America lies the Australian continent, where only twelve million people live in its three million square miles, and two and three quarter millions of them are concentrated in a narrow coastal belt, with a favoured climate, that extends north and south of Sydney. Most of these live in the Sydney metropolitan area, which has generated enough traffic to warrant a long, rapid transit commuter railway being built to serve the city and suburbs, additional to the existing suburban network. In the sparsely inhabited Australian continent, Melbourne has also decided on an underground railway as a means of freeing congested central streets from traffic congestion, and there has been mention of two other cities with like ideas.

In European Russia there are already five Metro systems serving major cities; and in Soviet territory east of the Urals, where development is probably more pronounced than anywhere else in the world, cities that were virtually only settlements fifty years ago may well follow the 'Metro' trend in European Russia, and themselves possess systems within the next ten to twenty years.

In the United States of America the pattern of development has stabilized and one might say that the USA is largely consolidating and expanding its industries. Its cities may not grow much larger, but their metropolitan regions are developing, and although in the past America has neglected mass public transportation (with some notable exceptions) the Federal Government is now actively encouraging it on a regional basis. There is an increasing awareness in America that however efficient the highway network around a city may be, it is likely to produce growing confusion where highways empty into a city centre, unless there is some form of mass transportation to take over there and relieve the flow.

The State of California, where abundant nature has produced a population explosion, is particularly interesting from a 'rapid transit' point of view. Within the State, two large complexes, Los Angeles and San Francisco, contain nearly twelve million people, about half of the total Californian population. Los Angeles and its environs constitute the largest built-up area in the world. (Viewed from above the Griffith Observatory there seems no end to it.) Public services generally are hard put to keep up with this fast-growing complex, where an extremely big and costly rapid transit project has been under consideration for years; but public reaction to it up to the present has not been sufficiently favourable to give it the go ahead. Meanwhile the city's public transport and its highly efficient road system are presumably able to cope. This situation is reflected in other growing American metropoli, but Los Angeles is referred to here principally for comparison with San Francisco and its Bay Area, which resemble it in growth potential.

The transportation problem in San Francisco was faced up to ten years ago with the decision to build an inter-county rail rapid transit system seventy-five miles long and to re-route the downtown portion of San Francisco's street car system underground as part of the same project, as well as to integrate new and existing bus services with rapid transit services. Unlike Los Angeles, which is accessible by land from many directions, San Francisco's business and social hub, a comparatively small area of high prestige value concentrated in the tip of the peninsula, is limited in its busiest approaches to two great road bridges and the Bay ferries, which it is imperative to relieve of overburdening traffic. A newly-built double-tracked Tube tunnel now lying beneath the Bay, through which rapid transit trains will run between San Francisco and the mainland, will accomplish this.

One might say in appraisal of the situation of the two cities that Los Angeles can just about afford to wait, perhaps to observe the operation of the Bay Area Rapid Transit project and profit thereby, but that San Francisco could not afford to wait, and in the event is acquiring the most advanced rapid transit system in the world.

The tide of rail rapid transit construction is slowly reaching more large cities in the USA, as will be seen in the main text of this book; and in Canada the future may reveal a similar picture. Toronto and Montreal have endured the inconvenience and upheaval inherent in the projection of underground railways through the heart of cities, but the Subway in one and the Metro in the other are now regarded as beneficial attributes, repaying their cost gradually but surely in the tangible form of enhanced land values, and intangibly in time saving and social gain.

In the Far East, Japan is an obvious candidate for the best means of mass transport available to move its teeming millions into and out of its astonishingly fast-growing cities. Several of the biggest are located along the Pacific coastal belt of lowland on Honshu Island (Japan's mainland), which is otherwise mainly mountainous. Until 1964 they were connected by the Japanese National Railways' narrow-gauge trunk railway, and since then also by the super-speed New Tokaido Line of standard gauge. This is planned eventually as an end-to-end trunk railway that will stretch unbroken for 1,200 miles from the northern Hokkaido Island, through a long undersea tunnel to Honshu and thence down to the southern tip of Kyushu Island.

Many Japanese cities were heavily damaged during World War II, but more or less ever since the arduous job of reconstruction was concluded, underground railway construction has been going on in the three biggest cities, and now Yokohama has also begun work on a four-line network totalling thirty-nine miles, which it is hoped will be completed by 1985. Although Japan is rapidly making good her pre-war deficiency in first-class roads, in railways it has always

22

had, and still maintains, complete confidence in their viability and efficiency. The intensive use of railways is nowhere more pronounced than in urban areas, which are threaded by many private lines as well as those of the JNR.

The policy now is to interwork their suburban services with those of the conventional rapid transit or subway lines, so that the big dormitory areas around at least four principal cities, served either by suburban trains or subway trains, will practically nowhere be out of reach of one service or the other. To make this possible, Japan has added in the seven years ending 1968 no less than forty-seven miles of new subway or rapid transit route to its existing systems, and as much again is in plan or actually under construction.

Finally there is the southern hemisphere, in which only about 20 per cent of the world's population live, and where at present the only integral rapid transit system is that in Buenos Aires. There are four other southern cities, additional to the Australian cities mentioned, which have tentative plans, but it is in the northern half of the world that most future underground railway activity will take place.

Airport Links

Apart from its function of providing mass transportation for cities, rapid transit has recently acquired an additional role, that of providing rail links between cities and their airports. A conventional rapid transit line with intermediate stations that links Cleveland, Ohio, with its airport, and is the first of its kind in America, was opened in 1969. Early reports quote a satisfactory number of air travellers using the line and no doubt its operation will attract the keen interest of other transit authorities.

Another American airport line, this one operating between New York and its Kennedy Airport, has been suggested, and there have been similar suggestions regarding London and its Heathrow Airport. In Kiev in the Ukraine the first section of a passenger monorail between the city and its Borispol Airport is reported to be going into service in 1970. On this monorail line a steel rail serves as a stator for the electric motor in the passenger car, the motor moving in a magnetic field and taking the car with it. (The Tokyo monorail cars are driven by conventional motors, but in Kiev they appear to be a linear inductive type.)

In all, it does seem that we shall be hearing more of monorail and conventional railways being built to connect airports to cities – although it is not inconceivable in the long run that vertical take-off and landing types of aircraft could alter the whole conception of large passenger airports necessarily located some distance from cities, and make such connecting railways unnecessary – at least for passenger traffic.

Historical development of rapid transit

Thus far the author has dealt briefly with current rapid transit activity around the world, but at this point he returns to the beginning of rapid transit in the nineteenth century. By referring to early developments that led to the various types of rail rapid transit in operation today, he hopes to provide a more comprehensive picture of the whole.

London pioneered urban underground railway construction in 1860, when work commenced on what became known as the Metropolitan Railway, a line in shallow tunnel built by the trench or cut and cover method, that ultimately connected the west of London with the City. It extended from Bishop's Road, Paddington to Farringdon Street, Clerkenwell, a run of about four miles, and was opened to passenger traffic in 1863. The terminal stations of three trunk railways lay along its route from west to east and the Metropolitan's eastern terminus was within the City.

Powers for its construction were obtained partly on the premise that it would help to relieve congestion in city streets crowded with horse-drawn traffic – and it probably did this for a time; but the railway was largely a commercial enterprise, proving in the event to be a financial success. The Metropolitan did in fact meet a demand for a swift means of transport along a well established traffic route, until then served only by one form of bulk passenger transport, the horse-drawn omnibus. This pioneer underground railway, worked by steam locomotives, thus came into being essentially to meet an existing demand, and only subsequently did it and its successors expand to *create* a demand, as instanced by the creation of 'Metroland' north-west of London. (Although subsequent ventures such as the District and Central London Railways owed their beginnings to existing demands, they too created demand by thrusting out into potential suburbs).

The adoption of electricity as motive power brought about the greatest change in urban railway history that has so far occurred. As much as it is concerned with underground railways, this event dates from the inauguration in 1890 of the first 'Tube', the City and South London Railway. Electricity as a clean and efficient means of propulsion permitted underground travel without the large-scale ventilation necessary for steam railways, and therefore made deep-level tunnelling, similar to that already successfully driven by the circular 'shield' method under the River Thames, practicable for underground passenger railways. From this period began the process of electrification, and deep-level tube construction, that has developed and ramified into the many Metro, Subway and Underground systems of today.

The history of the elevated type of urban railway is one of equally successful operation, but far less expansion. Elevated railway systems, as distinct from single elevated lines, developed principally in America. Of the two mentioned here, and referred to later in the main text, one in New York has virtually disappeared and the other, in Chicago, is to be so linked with the city's subway network as to lose its identity as a purely elevated system.

The first in New York dates from 1870, and like the Metropolitan Railway, it was built to meet an existing commuter demand. In the mid-nineteenth century most of the city's light industry moved from Manhattan to New Jersey on the mainland in search of local labour. After the completion of large housing projects in North Manhattan and neighbouring boroughs it moved back, creating a large commuting public in consequence. A system of elevated steam railways built to carry this traffic was subsequently electrified, and existed for thirty years before New Yorkers, conscious of their noisy 'El' and its obtrusive structures, turned in 1904 to the first of the Subways, and progressively demolished the 'El' as more subways were built.

In Chicago the elevated railway system, whose lines radiate from Chicago's famous 'Loop', still comprise most of the city's rapid transit rail system. Portions of it may disappear as a new master plan for Chicago's traffic develops, but most of it will presumably continue as an important part of the city's rapid transit network.

The elevated monorail system is of the same family, but with one or two isolated exceptions, it is a modern conception – those operating being confined to specialized services such as city–airport links and exhibition lines. As mentioned earlier, the monorail in Kiev is the most recently reported, but in Japan the Tokyo to Haneda Airport line of eight and a half miles has been functioning since 1964. It is served by units of three permanently-coupled cars with a capacity of 340 passengers per unit (about one-third that of a conventional six-car underground train). The line stops just short of Tokyo's central district, alongside a main-line station, and some passengers continue their journey by ordinary train. Construction costs for this type of railway are said to be about one-quarter of construction costs for an underground railway of equal length. Passenger totals on the Haneda line were initially disappointing, but have steadily improved in the last three years.

Summing up the potential of elevated railways in general, it seems unlikely that any more of conventional design will be built within central built-up city areas, mainly because underground railways offer a better alternative, presenting no surface structure to impede street traffic. In the case of two rapid transit systems still wholly or partly under construction, the San Francisco Bay Area system and the Rotterdam Metro, both feature imposing stretches of aerial (elevated) railway carrying twin tracks on reinforced concrete structures, supported by single pillars. The aerial structure, however, ceases outside at the approach to central areas, where the tracks descend and continue in tunnel under the centres.

One might say that every city or community has its own exclusive problem in endeavouring to keep street traffic on the move, and at the same time to mitigate the ill effects of its presence. Those built originally to a medieval plan, favoured by geography perhaps or by improved trade, and thus progressively attracting people and traffic, are faced with additional difficulties.

Most cities rate their central space at a high premium, and it is here that the effects of traffic congestion are most disrupting. Excluding obvious measures to accommodate the motor vehicle such as demolition of property to widen streets, or building raised motorways, to both of which there may be practical and aesthetic limits, segregation of public and private traffic has proved a positive step towards a solution. In the manner described here it constitutes a fourth form of developing rapid transit, and is simply that of putting public transport services underground in congested areas.

In some major European cities this is being done at the present time on an extensive scale, but there the traditional and principal means of public transport are tramways. In London, tracked trams and subsequently trackless trolley-buses were removed about fifteen years ago, partly because their services were disrupted wherever they shared busy thoroughfares with other traffic, and partly because they were thought to be one of the main causes of traffic congestion – although today there is some opinion that they might have been removed too hastily.

The European cities referred to were faced with the same decision as was London, the difference being that in the absence of any high-capacity rail network similar to the London Underground, these cities relied heavily on their tramway systems. In line with development in their respective cities, tram services were intensified and capacity increased with bigger and better vehicles; but as their routes inevitably converged on central areas, trams became trapped in general traffic queues and service delays multiplied as congestion increased.

The serious consequences of this deterioration in vital services were realized some years ago. Decisions were taken then not only to retain trams, at least for the time being, but to restore and increase their efficiency by building special tunnels and stations for them under busy streets. In most cases these underground ways were designed large enough to accommodate underground trains, should circumstances warrant adapting them to conventional U Bahn (underground railway) standards. These were major decisions, involving large capital outlays and much inconvenience to road users during the construction periods, but it is confidently expected that the ends will justify the means. Although in Budapest and in some American cities there have been tram subways in existence since the late nineteenth century, the projects in Frankfurt, Cologne, Stuttgart and elsewhere are the first instances of large-scale adoption of this form of rapid transit in urban areas.

What might be described as a fifth type of rapid transit is that which combines surface suburban railways with urban underground railways. As far back as 1870 or thereabout, a trunk line to carry both urban-suburban and main line traffic beneath London was suggested as part of a scheme to link the North of England with the Continent via a Channel Tunnel. Eighty years later another proposition for large-profile railway tunnels to be built beneath London was put forward as part of the recommendations in the London Plan Working Party Report. The tunnels would have taken both underground and suburban type trains.

These ideas, neither of which was taken up, are today turning to realities. In Kharkov in the Ukraine just such a project for a dual-purpose tunnel is under way at the present time. In Tokyo the principal of interworking urban and suburban traffic has already been adopted. In Paris a new variant, the East-West Regional Express line, twenty-eight and a half miles long, whose inner stretches are underground, is being built. Although part of the Metro system, it will absorb two former SNCF (State Railway) suburban lines, the Saint Germain and the Vincennes, which now terminate at the city's periphery. On completion of the new line, suburban passengers will be brought right into central Paris on Regional Express trains which will stop at certain underground stations for interchange with existing Metro services. All these schemes employ the principal of projecting suburban surface lines beyond their former termini, underground into city centres.

Planning and financing rapid transit.

The author has so far referred to rapid transit in terms mainly of accomplished projects, and turns now to the essential preliminaries preceding them. To confer a modern rapid transit system on a community calls for a division and sub-division of work demanding expertise at its highest level. As some say, it demands the expert knowledge of civil engineers who have applied themselves to the problems of mass movement in cities.

For any such Metro or Underground proposition which starts from scratch, so to speak, one would expect the preliminaries and the actual work of constructing and equipping a basic network of lines to take upwards of ten years, or even more depending on the network's extent; and for the project to call for the services of a wide range of specialists in their own fields, who would probably include statisticians and economists, as well as geologists, civil engineers and those equipped to deal with the intricacies of land and rights of way acquisition.

The wherewithal in terms of money and material are generally first considerations. Considerable capital, which is never to hand, must normally be obtained from the public – and in recent times from governments also. This situation has arisen following the gradual transfer over the years of most privately-run city transportation enterprises to municipal ownership and operation. It follows that if a new system of public transportation (in this case, rapid transit) confers an ultimate benefit on the community at large, the community must find the money, or a greater part of it, to pay for it; the methods employed to raise capital varying considerably.

For the extensive San Francisco project, needing a possible $1,170 millions for completion in its entirety, the sources include local taxation and Federal and State grants. Because of changed circumstances not easy to predict over a construction period of eight and a half years from January 1, 1963, the Bay Area's project's original estimated cost, $792 millions, which allowed for inflation at the rate of 3 per cent per annum, has been considerably exceeded.

For London's Victoria Line, provision for the estimated cost of

£56 millions was made by the Exchequer in the form of a loan, on which interest has to be paid by London Transport: although the calculated interest so far has been capitalized, and the whole business of providing capital for suggested new lines for London, and perhaps elsewhere in England, is a matter for future consideration. As with the San Francisco project, costs have exceeded the original estimate.

In Japan, over the present period of intense rapid transit construction, expenditure ran at a peak level in 1968. In the two largest cities, Tokyo and Osaka, the means of financing new construction takes the form of Treasury loans and public bond issues. The reported total cost of new subway construction for the Teito Authority in Tokyo, over the years 1962–9, is 349,100 millions of yen, or about £349 millions. For Osaka the total costs over the years 1965–75 are given as 375,464 millions of yen, or about £375 millions. (The equivalants are pre-devaluation). It will be seen from these examples, which reflect many others, that urban rapid transit is financed considerably from public sources, with state participation. The latter's participation increases where urban services of state railways become an integral part of rapid transit systems.

Physical aspects.

Turning to the physical side, planning of a system in some detail, allowing for subsequent modification, may precede arrangements for capital funding of the project. This was the case with London's Victoria Line, where it was necessary to be prepared for an immediate start on construction because of protracted delay in official authorization, which in view of the pressing and obvious need for the line, planners had to anticipate. Milan may be quoted as another instance. Here a plan for an underground railway was originally submitted by the Municipality to the Transport Ministry in 1953. Revisions of the original plan were necessary in the immediate ensuing years because of technical factors and the need to take into account town planning and development conditions arising meanwhile. Arrangements for the first major public financing of the re-constituted system were not instituted until 1957, actual construction started in June of the same year, and the first line was opened in 1964.

Some of the problems that confront planners have been concerned with ensuring the maximum benefit or usage of a rapid transit system based on future land development plans, or that created by the advent of the system itself. As an example, a new satellite community outside Frankfurt on Main, with 65,000 inhabitants, will be centred around a shopping area and will itself be surrounded by large employing agencies (works and factories). At the same time considerable traffic is expected between the community and the city, and two important routes of the new rapid transit system will eventually connect one with the other. In Washington, the 25-mile rail rapid transit scheme originally proposed has been so long in coming to fruition that its original concept no longer dovetails with development trends, or anticipated requirements. Part of the necessary re-thinking concerns probable expansion into a regional system and includes the re-alignment of a route in the original plan.

The importance of superimposing so permanent a feature upon a city's established pattern of lines of communication is apparent, when one considers that an underground railway network will be an arterial system of transport with the highest capacity, into which traffic will be fed from every other source, especially at its centre. Stations must be planned at points of main concentration and dispersal, and in inner areas their distance apart must be kept at a minimum where a maximum benefit is desired. This in part will depend on the land contour affecting gradients, the line's depth, the minimum distance commensurate with effective operation, and the public's reluctance to descend to great depth in order to travel short distances. Generally, the shallower the line the closer can be stations.

It would be impossible, for instance, to run a sub-surface line under part of San Francisco. The hilly district south of the waterfront at Fisherman's Wharf is a highly populous area of closely meshed streets that in a flatter district could qualify for an underground tram system. Up or down certain of these streets the cable cars operate on their switchback routes at gradients (up to 17 per cent) much too severe for any shallow tunnel route to operate with modern traction methods.

Nowadays a tube railway in London is preferably routed in as direct a line as is physically practicable, whereas in early construction it was better to follow the course of streets and avoid passing under private property. In this way the length of line was usually added to and sometimes awkward curves created, but at least there was less to pay for easement costs. The high cost of construction today makes this less of a consideration. Apart from the obvious advantages of a straighter run, such as greater permissible speeds, the line is that much shorter. This principle is followed in other countries also.

There are too many systems in too wide a field for individual examples of construction to be mentioned here, where ordinarily they would merit it. The San Francisco Bay Area project, however, not only dominates the field, but one of its many special features, the under-water crossing, deserves special mention. As previously referred to, the Trans-Bay Tube is designed to replace existing ferries and relieve the gross overloading of the Oakland Bay Bridge, and to a lesser extent, the Golden Gate Bridge. The Tube's tracks will carry most traffic since they will be the link between three radial lines on the mainland and San Francisco city.

There have been previous examples of underwater tunnelling similar to the Bay Tube, notably in Stockholm and Rotterdam, both adding to the sum of total knowledge of the subject. To extend Stockholm's 'T Bana' system a short crossing of a waterway was made in prefabricated concrete double-track tunnel. This was built in dry dock and floated to the site, to be sunk there to a depth of twenty-four feet. In Rotterdam the northern (city) portion of the Metro was partly constructed in prefabricated concrete tunnel sections which were floated to their respective sites along a canal, dug within the city for that purpose only and then filled in. Under the Meuse River the Metro line runs through double-track tunnel composed of twelve sections, each 280 feet long. These were sunk so as to rest on supporting piles, each pile having an adjustable

3. *Above:* Berlin (West): U Bahn station Alt-Mariendorf, Line 6; *left:* Frankfurt a-M: early construction work on U Bahn under Eschersheimer Landstrasse June 1963.

4. *Above:* Frankfurt a–M: U2 type trains on Nordweststadt–Hauptwache (AI) route; *below:* Hamburg: U Bahn train of DT3 type.

. *Above:* Munich: U Bahn two-coach train at Alte Heide; *below:*
Munich: U Bahn car interior.

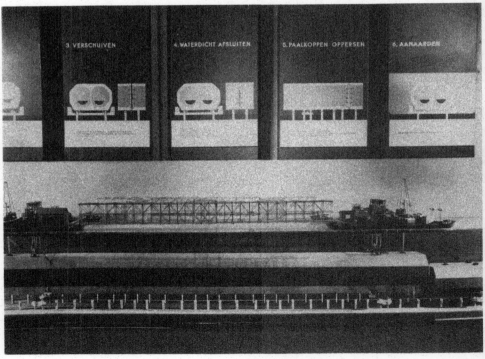

6. *Above:* Rotterdam: Metro train; *below:* Rotterdam: model of River Meuse tunnel section being sunk to river bed.

head. Injections of grout under pressure forced the pileheads upward against the tunnel bottom until it rested evenly on all the
piles. The individual sections, with rubber profiling at each end,
were forced together by hydrostatic pressure to make the whole
watertight. The tunnel lies two metres below the river bed, causing
no interference with the passage of shipping.

The San Francisco Tube employs similar techniques, but on a
vastly bigger scale and with more specialized procedure. Its fifty-
seven binocular-shaped sections have been laid in a trench sixty
feet wide, dredged in the Bay floor, between two ventilation buildings, the distance between being 3·6 miles. The steel shells of each
section, fabricated on the Bethlehem Steel Corporation's 'ship-
ways', are 47 ft 10 in in overall width, 21 ft 6 in high, vary in length
from 273 to 366 ft, and weigh up to 12,000 tons each. They include
sections with horizontal curves, vertical curves, and two with both
horizontal and vertical curves, to enable the Tube to follow sea-bed
contours and the Tube's alignment. Their end profiles show side-
by-side running tubes each 17 ft in diameter, separated by two
rectangular openings, one above the other. The upper is the ex-
haust air duct and the lower a gallery for equipment.

Although the Tube's alignment avoids the epicentre of recorded
large earthquakes, the Tube itself could be affected by ground
deformation caused by such disturbances. (The 1906 San Francisco
earthquake was caused by slippage along a local fault in the earth's
surface, and the Bay lies between two such faults.) Following a
geological survey, therefore, and extensive tests over several years
of the shock-absorbing qualities of soft soils overlaying the bedrock,
the Tube's alignment was decided along a course, determined by
the approaches to San Francisco and Oakland, where possible
shocks would be most effectively absorbed. In addition, structural
features in the Tube itself afford protection from earthquake
effects. They include considerable stiffening of the sections' steel
shells with longitudinal and transverse bracing against buckling.

The Tube ends terminate in ventilation buildings, where the
Tube joins with tunnel sections running beneath San Francisco and
Oakland. The building on the San Francisco side is actually a
caisson-like structure 100 feet high, standing almost wholly

submerged, just offshore. Within the building, tunnel joints had to be both watertight and flexible, the latter to accommodate any earthquake displacement between the buildings, the Tube and the tunnels. The terminal joints allow for possible movement in both horizontal and vertical direction. Placing the Tube sections in position was a similar operation to that performed in Rotterdam, in so far as the sections, with temporary bulkheads, were launched, floated to their respective positions and sunk by filling ballast pockets on top of each section.

The Tube now lies, covered by protective layers of gravel, sand and stone along its entire length, at depths varying with the contour of the Bay bottom, the lowest point being about 125 feet below sea level.

PART II

REVIEW OF WORLD RAPID TRANSIT SYSTEMS

In the following pages will be found brief descriptions of the many rapid transit systems operating or under construction, and the city background in which each operates or is being built.

VIENNA

The walls of a second-century Roman camp, enclosed by later ramparts and bastions, imposed a concentric pattern of streets on the centre of Vienna which are about the only reminder that this was once a heavily defended city. A visitor today penetrates the outer Gurtel (girdle) and the inner Ringstrasse either to look at the city's splendid buildings or its inviting shops – or just to saunter along its crowded streets. The impression in Vienna at the present time is one of prosperity, and as such is a true reflection of the country's well-being.

The rise in car ownership in Austria from fifty-eight per thousand in 1960 to 140 per thousand in 1968 explains the crowded central core of streets, where trams have never penetrated. But Vienna is still essentially a 'tramway' city, developing from the horse-drawn variety, through steam locomotive-hauled vehicles to electrified tramways introduced about the year 1900. For over half a century trams have carried by far the greatest proportion of traffic on the city's public transport system. Even as late as 1967, when the grand total for the Stadtbahn (town railway) and buses and trams had decreased slightly to 434 million passengers, the trams' share of that total was 308 millions, or 71 per cent and their lines accounted for 213 miles of the total public transport network of 384 miles.

Under a traffic segregation scheme, various underpasses have been built for trams and pedestrians where traffic on radial routes crosses that on circumferential routes. This has since developed further by the construction of sub-surface tunnels, and halts or stations, for trams. Four main tram routes were put underground on the Ring Line (the inner circular route) in 1966, and in 1969 trams were diverted into a tunnel built beneath a southern portion of the Gurtel, and extended a short distance as spur tunnels on either side. These undertakings warrant mention in this book since the tunnels are suitable for adaptation for, and may form part of a future conventional underground system. Preliminary excavation has, in fact, commenced on Line 1 of a U Bahn network for Vienna, which will have approximately six miles of route and is scheduled for completion in 1980.

The Stadtbahn, which was inaugurated in 1898 as a steam line, leans more towards a tramway system than an underground railway – although it operates in its own rights of way, and for nearly four miles of its total length of sixteen and a half miles runs in tunnel. Roughly encircling the central core of Vienna, it follows closely the Wien river from west to east, then the Danube Canal north-westward, and finally the Gurtel southward. There are actually four operating lines (DG, GD, WD and G), which are physically connected, and twenty-five stations. The system was electrified in 1925. There is a project under consideration to extend this system by a north-easterly spur passing under the Danube river to link Vienna's growing suburbs in the north-east by rail to the city centre.

ADDITIONAL DETAIL

Authority: Wiener Stadtwerke Verkehrsbetriebe, which also operates the trams and buses.

Passengers carried in 1969: 74,500,000.

Tunnel construction: The tunnels are double-tracked, generally of cut and cover construction with vertical walls and elliptical roofs, but in places are of circular cross-section.

Traction current: Obtained from the Vienna Electricity Works, transformed to 750v DC through sub-stations, and fed to the cars through overhead pantographs.

Rolling stock: This consists of motor-cars and trailers of tramway type, twin-axled, with centre-buffer automatic coupling to form trains of up to nine cars.

Signalling: The AC track circuited automatic block system has two-aspect colour light signals.

Vienna is also served by the Schnellbahn (rapid transit railway) operated by the Austrian Federal Railways. This system starts in south-west Vienna at Lessing and runs north-east, interconnecting with the Stadtbahn at Landstrasse station. It then continues, to cross the Danube, and then forks to Strebersdorf and Sussenbrunn respectively. It has a common fare tariff with the Stadtbahn.

BRUSSELS

Brussels, capital of Belgium, has experienced a considerable building boom since it became the meeting place for international organizations such as the European Economic Community (Common Market) the European Atomic Energy Community and others. Many foreign business concerns interested in European trade have also established their headquarters in this city of about one million people. The consequent influx of permanent officials and trade representatives has helped to bring prosperity to the city, but it also contributed to the congestion of traffic on streets, causing a traffic problem which has been tackled by road improvement works, and also by providing trams, or street-cars, with their own underground rights of way in central areas.

Trams are by far the largest of Brussel's public transport carriers. The total length of all their routes exceeds 120 miles (196 km.) and they carry nearly three times the number of passengers than do the city's buses. Of their new tunnel lines, some three and a half km in central Brussels, with six underground stations, is already open, and a further two km will come into service in 1970. Thenceforth this inner network of tunnel lines will be expanded until it covers forty km of route.

The process of putting trams underground, where they operate under a system of automatic signalling and braking, is a prelude to the establishment of a conventional Metro system in Brussels. (It has in fact been given the name 'Pre-Metro'). The tram tunnels and stations are so constructed as to be adapted to take conventional Metro trains; and with the advent of the latter, expected some time in 1972, the trolley busses now operating will progressively disappear.

PRAGUE

The centre of Prague, Czechoslovakia's ancient capital, has long been subjected to a degree of street congestion that, whilst it affects all traffic, particularly disrupts the public transport services. Prague is famed for the richness and variety of the old buildings and monuments at its centre, but this historical area is ill adapted to deal with the pressures of modern traffic. Its streets, some of which carry a mixture of private transport and public transport in the form of buses and street cars, quickly become congested. The situation seems unlikely to improve whilst surface traffic of all kinds has access to the city and is not segregated one type from another.

A partial solution appeared to lie in the separation of street cars from general traffic by diverting them underground in the central area, and some preliminary work on this project, which envisaged three street-car lines running in 14·3 km of tunnel, had in fact been done by 1967. The original concept of a transitory underground street car system, however, had meanwhile been altered to a perspective underground railway system, and at the time of writing (1970) work on this newer project is actively under way.

The change of plan came about as a result of a decision that Prague's future transportation needs could be adequately met only by a system of underground railways. The view is that in the long term such a system will solve the problem of traffic congestion, but there are interrelated and earlier objectives. For instance, new housing projects at North Town and South Town, being developed to catch up with Prague's general housing shortage, are considered to warrant mass transit facilities. These will be met by a transverse underground line that will also go some way to relieving congestion at the city centre.

The new concept, which will probably be known as the Prague Metro, is for a standard 4 ft 8½ in (1·435 m) gauge system of three transverse lines, A, B and C, and a supplementary radial line D, totalling in all approximately 45 miles (73 km) of route, of which about 35½ miles (57 km) would be underground. The transverse lines where they meet at the city centre would form a triangle. The work at present in progress is on line C, that which will link the housing development areas to the north and south of Prague, mentioned earlier, to the city centre. The work is described as the first stage of Phase 1 of the whole scheme, which comprises 18·65 miles (30 km) of railway route, including 12·4 miles (20 km) in tunnel, and will continue until 1980. Stage 1 comprises 4·2 miles (6·8 km) of route, all of which will be in tunnel except for the 480 metre Nusel Bridge, which crosses the valley of the Botic River in southern Prague.

The tunnels, which for 2·9 km will be rectangular, built by the 'Milan' method of cut and cover, will for the remaining 3·4 km be in circular tunnel, lined for the most part with concrete segments, giving an internal diameter of 18 ft 8 in (5·1 m). They will lie at an average depth of 13 metres, with a maximum depth of 23 metres, and are being built in, or driven through soil that ranges from sand and gravel to shales and quartzite. An excavating shield from the Soviet Union is being used in tunnel construction.

There will be nine stations on Line C, eight of which will be underground and one situated in the abutment of Nusel Bridge. Seven of these stations will have island-type platforms and two will have side platforms. The railway cars will be built in the CKD Prague works, and will have a capacity of 300 passengers per two-car unit normally and 420 at maximum. Trains will be made up of two cars, which will operate at a maximum frequency of forty trains per hour.

Prague is at present served by a system of buses, trolley-buses and street cars. The total of all their routes amounts to 486 km, of which the street car route share is 131 km; but it is reasonable to suppose that street car services will be withdrawn wherever their routes are adequately served by the future underground railway system.

PARIS

To the world in general, Paris is perhaps best known as a city of culture and fashion, with wide streets and avenues and elegantly façaded buildings. It is probably not so well known that Paris, together with its district, is also a very big industrial, commercial and suburban complex, several times larger than any other French city complex, and that the concentration of people and traffic at its centre has reached such proportions as to necessitate the whole Paris Region being considered for re-planning.

Nothing like this has happened since the late nineteenth century, when much of central Paris was destroyed in a civil insurrection in 1871 and had subsequently to be rebuilt. It was Baron Hausmann who, in his redevelopment scheme for Paris, created the famous Grand Boulevards and connecting avenues that radiate from the centre to the gates of the old city. Today the Paris Region contains nearly nine million people within its 100 square kilometres. The severe congestion in its central *arrondissments* (municipal areas) has been attributed partly to the traffic they generate, partly to the failure of public transport to keep pace with the growing population spreading over a large area, and partly because, it is reported, there are so few dwellings in the inner areas that possess garages, resulting in an estimated 20 per cent of total road space being occupied by parked cars.

Whatever the causes, the effects have been such as to warrant the creation in 1960 of a planning organization for a greater Paris Region, whose objects include the development of approximately 1200 square km into an area suitable for housing, employment and recreation. This embraces a plan for drastic decentralization, the result of which will be the creation of eight new urban centres each of one half to one million population. These will be located along a 'corridor' extending westward from Paris along the river Seine valley towards Rouen, and will be linked to it by new motorways and an express Metro. New communities to the north and south of Paris will also be linked by fast road and rail services.

These developments, together with some considerable rebuilding in parts of central Paris, are expected to occupy the next twenty to thirty years, but construction of the new express Metro is already well advanced.

Geographically, Paris resembles London on a smaller scale, with its river curving through the city like the Thames, and its main railway Termini stopping short of the centre. Like many another city in the mid-nineteenth century, the exodus of much of its working class population from city homes to more pleasant outer suburbs created a commuting public – its work places remaining in the city; and it was this migratory move that gave rise to a demand for the then new means of mass transportation, suburban railways. These circumstances might have resulted in a radial system of steam-operated suburban lines, or in a circular system connecting the main-line termini, but the consequent expense and constructional difficulties of superimposing either on the city's street system delayed any action for so long that the new medium, electric traction, came into being meanwhile and made practicable the urban underground type of railway. Paris, unlike New York, was thus spared the putting up and subsequent pulling down of elevated urban railway lines, which however elegantly designed (and there were some elegant designs) would hardly have blended in with the style of Parisian architecture.

In July 1900, the first Metro line (No 1) was opened. It ran east-west beneath central Paris between Porte De Vincennes and Porte Maillot, and with two short branches opened a few months later proved its popularity by carrying, over its 13·35 km of route, nearly 18 million passengers in five months. The northern half of a circular line was next added and the circle was completed, together with several radial lines, over the following ten-year period. Only during the first and second World Wars was this steady expansion curtailed – and now, in 1969, the Metro, which today includes the Sceaux suburban line, totals about 127 miles (205 km) of route, comprising thirteen radial lines, two circumferential lines and the Sceaux line.

The system is so much a close-meshed network of lines, apparently illogically multi-directional but severely dictated by topography and passenger demand, that a detailed description would run into many pages. It is best to describe the Metro as a very convenient means of travelling short distances, mainly because the stations are close together and generally not far below ground. It has one unique feature (there is a first-class coach in the middle of the train), and some features which were unique but have now been adopted elsewhere in the world, including its characteristic dome-shaped tunnels and its pneumatic-tyred trains. In spite of the intricacy of its network, it is reputed harder to get lost on the Paris Metro than on London's Underground, for instance; although both Authorities, realizing that underground travellers are specially prone to disorientation, have done their best to make direction as clear as possible. Finally, one can travel short or long distances on the Metro for the same basic fare, the only variation being in the first and second-class tickets and on journeys on the Sceaux line, where one pays according to the distance travelled.

The Metro's pneumatic-tyred trains were, when first introduced into passenger service on Line 11 (after extensive testing of a prototype) a radical departure from the conventional underground train. The reasons for their adoption included quieter and smoother running than with steel-tyred trains, their increased adhesion to the track assisting in braking and acceleration – important considerations on the Metro with its closely spaced stations – and a reduction in weight of parts, which need not be so sturdy, as they are cushioned against shocks by the pneumatic tyres.

The rubber tyres are similar to those used on heavy lorries. Each bogie (truck) carries four of these rubber-tyred wheels, which normally run on specially broad concrete beam or steel tracks located on each side of a conventional steel running track. On the same axles are mounted doubling steel wheels with deeper than usual flanges. These wheels are normally held clear of the steel track, the rubber-tyred wheels supporting the train. Should the tyres become deflated, thus lowering the train, the steel wheels bear on the steel track and thus assume the load.

Two lateral guide bars are located outside the running tracks, but at a higher level, and upon these bear the smaller rubber-tyred guide wheels, one mounted at each corner of the bogie. At junctions, where a necessary gap occurs in the guide rails, the flanges of the steel wheels make contact with the steel running track and so guide the bogie. Positive traction current is collected through shoes in contact with the guide rails, and negative current returns via contacting shoes through the conventional steel track, which also carries the track circuit, or signal current. About one-sixth of the Metro's total passenger stock has pneumatic tyres (excluding that of the Sceaux line, which is built to main-line standards).

The new east–west Regional Express line mentioned earlier, whose tunnel sections have been driven deep beneath central Paris, might well be called the *piéce de résistance* of all Metro construction. About 28½ miles (46 km) long, it reaches from Boissy St Leger in the east to St Germain-en-Laye in the west. Its two main functions, when the line is complete, will be to link the outer suburbs and some of the newly created communities mentioned earlier with central Paris, and to absorb two French National Railway lines so that passengers who previously detrained at their east and west termini may be able to continue their journeys to the city centre and beyond, without a change of trains. The beneficial effects of this will be to provide suburban dwellers with a fast and high-capacity rail service into town, and to ease traffic conditions on the most heavily used Metro lines. Nine miles of the new line is underground, and parallels for some distance the No 1 Metro line, with which there will be interchange.

The tunnel sections, plunging deep beneath central Paris to avoid the many services and the Metro tunnels, lie at an average depth of 82 ft (25 m) below surface level. Unlike the tunnelling for London's Victoria Line, which consisted mainly of driving deep-level tunnels without disturbing the ground above, the Paris project included provision for underground complexes, containing for instance motor routes and car parks, necessitating in places excavation of huge trenches. In one case the space above the express tunnels houses two motorway tunnels, themselves flanking an underground ticket hall which in turn lies beneath a sub-surface concourse. Between the future Etoile and Auber stations there are no less than six levels, one above the other on top of the express tunnels, for underground garages and car parks. The express tunnels are big in themselves, in the box sections (with arched roof) and in the circular sections, being approximately 28½ ft (8·70m) in internal width and in diameter respectively.

The express Regional trains using these tunnels will only stop at a few important stations where the line parallels that of the Metro. During busy periods, nine-coach express trains, twice as long as ordinary Metro trains and capable of carrying 2,500 passengers, will run at a working speed of 37 mph and at frequencies of one train every 2½ minutes. The Eastern branch of the Regional Express line, extending from Nation to Boissy-St-Léger, as well as a section of the Western branch, between Etoile and La Défense, were opened for passenger traffic in late 1969 and early 1970 respectively.

The existing Metro is also planned for expansion, in one instance as an extension to link the surface railways in north and west Paris. Together with future Regional lines, the enlarged Metro and the new motorways will, it is hoped, be able successfully to cope with a 'Greater Paris' population of eleven millions in 1975 and perhaps twelve to fourteen millions in the year 2000.

ADDITIONAL DATA

Authority : Autonome des Transports Parisiens (RATP).

Number of stations : Urban system 266; Sceaux line 29; Regional Express line; 13.

Average distance between stations : Urban system 0·32 miles (0·52 km).

Passengers carried in 1969 : 1,277 millions approximately on the three systems.

Maximum service frequency : 8 to 61 trains per hour (on Lines 7 and 1 respectively).

Type of tunnel : (Metro) Double track of elliptical section, 23 ft 3½ in (7·10 m) wide and 17 ft 1 in (5·20 m) high.

Traction voltage and collection : The present 600v DC is being increased to 750v. On conventional stock lines, current is collected from a third rail (on lines with pneumatic-tyred stock, from insulated lateral guide bars). The Sceaux line operates at 1,500v DC, collecting current from overhead wires.

Number of coaches per train (Metro): 4, 5 or 6 according to line.

Average service speed including stops: 14·3 to 17·4 mph (23 to 28 km/h). Sceaux line 21·1 mph (34 km/h).

Signalling: Automatic block with three-aspect colour-light signals.

7. *Above:* Budapest: concourse under Blaha Lujza Square: *below:*
Budapest: tunnel showing flexible concrete segment construction.

8. *Above*: Milan: Rolling stock temporarily servicing Line 2; *below*: Rome twin-car unit in Garbatella station.

9. *Above :* Oslo: mid-city traffic scene; *below :* Oslo: train on the Østensjø
line, Eastern Rapid Transit System.

10. *Above:* Lisbon: Soldana Metro station; *below:* Lisbon: interior of Metro car.

BERLIN (EAST)

All that part of Berlin and its suburbs lying east of the demarcation boundary is the capital of the German Democratic Republic, whose population in 1968 was 1,070,000. In that year the author visited this part of the city, entering by way of the U Bahn No 6 line and Friedrichstrasse station, the only U Bahn station that is an entry and exit point between the east and west sectors. To the casual visitor the underground journey is strange, passing non-stop through darkened stations located in DDR territory – but it has long been accepted by regular travellers and there is no indication at present (late 1969) of any change.

Above-ground, the wide streets of the central areas are in distinct contrast to those in most other major cities. One expects and usually finds the latters' streets overcrowded and noisy with traffic, especially in downtown areas. In the central streets of the DDR capital today, traffic is brisk, but only moderate in comparison; a bearable and even pleasurable experience. Public transport vehicles have generally seen many years' service, but they seem adequate in number and certainly the services are reliable. Replacements are thought to be a question of priorities, as an important part of the country's policy in the last two decades has been to expand trade with foreign countries – and it has done this consistently year by year with evident benefit at home. Much new housing, in the form of apartment blocks set in green surrounds, has been built to the south-east of the spacious Alexanderplatz, providing new accommodation for people whose outdated homes in the inner areas were pulled down. Reconstruction on the vacated land is very considerable by any standard – and whilst at present only a moderate flow of visitors comes from the west, this part of Berlin evidently attracts numbers from the countryside and from eastern Europe generally. Reasonably there will be more when this large area is developed as projected.

D

Given these circumstances, the existing public transportation will need to be greatly augmented in future, and plans are under consideration for a new network of public transportation for Berlin, DDR. They are dependent naturally on available finance, but their aim will be to provide the best means of public transport for the expected population expansion and the changing pattern of land use.

In the meantime, additional to the trams, buses and trolley-buses, the BVG Authority in Liebnechstrasse (the West Berlin Authority is in Potsdamerstrasse) operates two of the city's U Bahn lines. These are Line A, of small profile, from Pankow to Thalmannplatz, 4·66 miles (7·49 km); and Line E, of large profile, from Alexanderplatz to Friedrichfelde, 4·41 miles (7·095 km).

The former line is about three-quarters in tunnel and the latter is wholly in tunnel. The two lines have twenty-two operative stations, and of the approximate 370 million passengers carried on the whole DDR system in 1967, approximately 74 millions used the U Bahn. A 1·2 km extension of Line E southward from Fredrich-felde is scheduled for opening in 1971.

For maintenance the U Bahn cars are conveyed on specially adapted rail wagons to the German State Railway's repair works at Schöneweide, about fifteen kilometres south of central Berlin. The elevated city railway, the Stadtbahn, also plays an important part in the city's transport system.

There are basic fare structures on both east and west U Bahn systems. That in Berlin, DDR is 20 pf for any distance (not transferable) and 10 pf for children and the medically unfit.

BERLIN (WEST)

Berlin's metro system, one of the oldest on the continent of Europe, was of course designed to serve a unified city. Before World War II its finely balanced services converged on the area around the mile-long Unter den Linden and Friedrichstrasse which crosses it, the area containing both magnificent palaces and public buildings and the principal shopping streets in Berlin.

This comparatively young city grew and finally dominated the northern European plain chiefly by virtue of its geographical position. The island in the River Spree at the old city's centre made a suitable crossing place for trade routes centuries before the advent of railways, and subsequently canal builders took advantage of waterway courses flowing conveniently east–west to put Berlin on the canal map, thereby increasing its trade and importance. Then in the mid-nineteenth century, railways were built to duplicate the road and water routes, and eventually multiplied to the extent that the city now has twelve lines radiating from it, the most important being on the long-distance trans-continental routes. From 1871 to 1945 Berlin was the capital of the German Reich, increasingly attracting to itself manufactures and commerce during all but the last years of that era. Concurrent with these activities came a growth of population. Numbers increased from 836,000 in 1871 to 2,000,000 in 1905, and then after extension of boundaries to form a Greater Berlin, to 4,350,000 in 1939.

In August 1961 the political barrier was set up which physically divided the city in two. The country and the railway systems were already under different governments and administrations, but the barrier also extended underground, causing the closure of several U Bahn stations on lines that originate or terminate in the western sectors of Berlin. These stations are those situated in the eastern part of the city. The Berlin U Bahn is today two separate underground systems operating through their own organizations. The descriptive matter which follows applies to that part operating in West Berlin under the Authority BVG (Berliner Verhehrs Betriebe), which also operates its tram, bus and trolley-bus services in West Berlin, but the physical properties of lines existent before the division apply, of course, to both systems.

Until recent years the BVG lines in West Berlin were known by the letters A B C, etc., but now they are numbered. There are two tunnel gauges and two sizes of rolling stock on the system. The 'Large-profile' lines are those running generally north–south (Nos 6, 7, 8 and 9.) and the 'small profile lines those running east–west (Lines 1, 2, 3, 4 and 5). The first to operate was one of the latter (in 1902) on an east–west section of line from Warschauer Brücke to Potsdamer Platz, 11·2 km long, built partly underground, but mostly along a viaduct. All of the lines built subsequent to the original line are in tunnel, with the exception of a few stretches in open cutting.

Since 1961 an entirely new line, first known as Line G and now Line 9, has been in operation. Besides serving new areas it has shortened and quickened journeys for passengers on the Tegel line (No 6) bound for destinations in the west. Previously such passengers had to continue into the city centre (thereby passing into and out of the eastern sector) before changing on to a westbound line. The No 9 line provides interchange at two points with east–west lines. Trains from Tegel, in the north, to Alt-Mariendorf in the south pass non-stop through stations in the eastern sector, with the exception of a short stop at Bahnhof Friedrichstrasse, which is a control point.

Extensions to the BVG system in West Berlin continue, and include an extension of Line 7 to the south-east, where at Britz Rudow, Berlin's largest housing project to accommodate 50,000 people is being built. Berlin is also served by an urban and suburban railway system and by a 5·8 km underground section of railway running north–south through the city centre, connecting main line railway termini.

ADDITIONAL DETAIL

The whole system has a route length of 59·5 miles (95·8 km) of which 46·6 miles (75·0 km) is in tunnel. 44·4 miles of the total route length is in the western sectors of Berlin. There are eleven lines, of which nine are in the west.

Stations : There are 116 in all (83 in the west). The oldest, elevated and part of elevated railway structure, are reminiscent of Earls Court Station (or King's Cross Station on small scale) in London, with curved iron and glass roofs over the tracks and the island-type platforms. The underground stations are generally rectangular in profile with pillars supporting the roofs. The tracks are standard gauge.

Rolling stock : There are 608 motored cars and 96 trailer cars (1969) on the BVG West Berlin system. The latest large profile cars are 50 ft 10 in (15·5m) long and 8 ft 8 in (2·65 m) wide. The latest small profile cars are 41 ft (12·5 m) long and 7 ft 7 in (2·30 m) wide.

Type of tunnel : Nearly all the tunnels are constructed just below surface level and were built in rectangular section, double-tracked, by the cut and cover method.

Signalling and traction current : Basically the signalling system operates on the automatic block principle with colour-light signals, but an automatic system is under experiment whereby signals are picked up by the train through induction coils. Traction current at 780 volts DC is collected through a third rail.

The BVG system in West Berlin carried 782,670,000 passengers in the year 1968, on all its public transport. 216,700,000 were carried on the U Bahn operating in the western sectors.

FRANKFURT ON MAIN

Frankfurt on Main is one of those European cities that is an important centre of road, rail and air communications. It has, however, a certain uniqueness among cities with comparably long histories, in so far as when its medieval centre was destroyed by bombing in World War II, there followed an intense, almost fanatical ten-year period of reconstruction that has transformed Frankfurt into virtually a modern city. Business life restarted and expanded and today the city is one of the largest and most highly commercialized in West Germany. Its bustling go-ahead characteristics, its big department stores and some of the goods they sell may be attributable to American influence, but Frankfurt still remains basically a very Germany city.

Viewed by day from the top of the 384-foot Henninger Tower in Sachsenhausen, a monument to the city's brewing industry, Frankfurt's city blocks appear angular and modern, but their outlines fade at dusk into a panorama of twinkling lights fourteen miles wide, and there are hundreds of elegant old buildings spaced about to make it still a most attractive city by day or by night. Frankfurt's remarkable growth in recent years is illustrated by the population figures: 1939, 553,000; 1946, 426,000; 1967, approximately 700,000. Modern Greater Frankfurt has expanded to treble the size of its city area in 1880, and its present traffic figures are even more impressive. Each morning some 500,000 passengers arrive at the Hauptbahnhof, Frankfurt's principal railway station, where there are 1200 train movements a day. At the airport there are about 1300 flights per week and the biggest road traffic junction in Germany lies just south of the city. But like most large German cities, Frankfurt suffers from acute traffic congestion at its centre. Its trams, numerous and heavily relied upon, follow their fixed paths and considerably aggravate the situation – or at least that was the position until October 4, 1968.

54

On that day the pressure was relieved to the extent of three services of trams, whose vehicles carried their first passengers beneath the central streets – and since Frankfurt was the first city in recent times to operate trams at sub-surface level, through tunnels that were specifically built large enough to accommodate conventional underground trains, the author devotes a little more space to the city and its new U Bahn project. Construction work began above and below ground as early as 1963, but in reviewing the whole finished project (which in itself is only Phase 1 of a vast road-rail improvement scheme) the work accomplished in five years is remarkable. In essence, more than half of a busy five-mile tram route, leading north from the city centre to Hohemark and Bad Homburg, has been re-aligned into its own, exclusive right-of-way and the rest has been diverted into tunnel. The latter, double-tracked, is two miles (3·2 km) long, and the track realigned on the surface, as far as Eschersheim, is 2½ miles (4 km) long. Motor traffic now runs virtually unimpeded along the busy main road, on each side of an enclosed median strip for trams, and such is the improved speed of the latter that they often keep pace with the automobiles above ground and outstrip them, mile for mile, through the tunnel section.

In anticipation of the U Bahn advent, a satellite town for 60,000 inhabitants, Nordweststadt, has come into being about ten kilometres north-west of Frankfurt's centre. This community has its own shopping and cultural centre, but being within the Greater Frankfurt area it naturally demands a good system of communication between it and the centre. This has been accomplished by connecting it to the new U Bahn (by a branch line) and providing the whole Nordweststadt-Frankfurt route with a completely new type of vehicle, an articulated twin-car unit coupled to form a train of four units (maximum), with automatically operated double doors. The units in fact form a conventional underground train, differing only from the majority of such types in their method of collecting traction current – from overhead wires through a pantograph. The ordinary existing tram cars, which interwork with the new trains, also have pantographs and sufficient headroom was allowed for them in the tunnel construction.

55

There are fixed halts with raised platforms on the open stretch of new line and also five level crossings for other traffic, controlled by traffic lights actuated by the passage of trams and trains. On the tunnel section there are five stations with side platforms, the principal one being Hauptwache (this is named after Frankfurt's chief guard-house which stood on the site. A replica has been built on more or less the same spot). This station merits a short description as it introduces features quite new to Frankfurt. Centrally situated, it lies beneath a spacious road junction, busy enough already to imagine it as a future Piccadilly Circus of Frankfurt – and more so when the station becomes an inter-change for another underground railway.

Associated with the U Bahn project is also the 'S' Bahn (suburban railway) project. This line will be in tunnel and will connect Frankfurt's main west and south Federal Railway stations. There is a considerable daily total of commuters and others between the two stations who at present use surface transport, and by providing an underground linking route for suburban train services the central streets will be relieved of that much traffic. The projected line runs eastward under the city before swinging south to cross the river Main. One of the underground stations will be at Hauptwache, which has therefore been built on three levels. There is a large concourse immediately below surface level, the S Bahn tunnels running east–west at the next level, reached by escalators, and the U Bahn at the lowest level, reached by a further bank of escalators, running north–south. One of the station's features is the group of shops and restaurant in the concourse.

In April, 1968, the author entered the S Bahn tunnels at Hauptwache when visiting the then uncompleted station, but in the following October they had disappeared. In fact they have been sealed off with wall panels and their existence will no doubt come as a surprise when the S Bahn is eventually completed and the station tunnels revealed.

Frankfurt's U Bahn network is planned for considerable expansion in future years. The existing north–south tunnel is to be extended southward and several more tunnel routes are to be constructed, radiating from the city centre, surfacing in the inner suburbs and continuing over existing surface routes to reach an eventual total route length of 80 miles (132 km), 20 miles (32 km) of which will be underground. Work is progressing, for instance, on a new route to Nordweststadt through Bockenheim, relieving the first line and providing a more direct city route. The existing trams will only be gradually replaced over the years by the conventional U Bahn trains.

There is no doubt that the U Bahn is proving popular with the travelling public. The new 'U2' type trains are fast and comfortable, and furthermore, passengers on both the new trains and older trams are not cut off from the passing scene until the line dips for the short journey through the tunnel. Finally they are usually carried to within a short distance of their ultimate destination, and have only to walk through a short pedestrian tunnel to reach either the sidewalk, the shops or their homes.

ADDITIONAL DETAIL

Tunnel construction: Double-track rectangular section with supporting pillars 15 ft 9 in (4·50 m) high, 23 ft 4 in (7·10 m) wide, mostly at sub-surface level.

Depth at Hauptwache Station: 64 ft (20 m).

Stations: Those existing have side platforms 312 ft (95 m) long. Average distance between stations, 0·37 miles (0·62 km).

Track gauge: 4 ft 8½ in (1·435 m).

Rolling stock, Type U2: Number of cars owned (1968), 32. Width of car 8 ft 8½ in (2·65 m); length of twin-car unit, 75 ft 5in (23 m). Two 150 kw motors per unit, equalling 204 hp. Four double doors on each side of each unit, automatically operated. Passenger capacity 230 per unit, including 64 seated. Maximum service speed 40 km/h. Maximum speed 80 km/h.

Number of existing tramcars on Frankfurt Strassenbahn street railway system: 268.

Total length of Strassenbahn routes: 911 km.

Total number of tram passengers in 1966: 180 million.

HAMBURG

Hamburg, the second largest city and the largest port in the German Federal Republic, lies about fifty-five miles inland from the North Sea on the River Elbe. In recent history the city suffered two major reverses, the first by war damage when, because of its strategical position and importance as a port it was one of the principal targets for Allied bombers in World War II, and the second when it experienced loss of trade through being cut off from East Germany (whose border is not far away) by political division. But Hamburg has shared in the remarkable economic recovery of the Federal Republic generally, and considerable extensions to its U Bahn and S Bahn systems, to handle progressive increases in its travelling public, serve to indicate its growing importance today.

Hamburg's urban traffic routes do not radiate directly from the city centre as is usual elsewhere. They are dictated by the presence of the Alster Lake, two miles long and half a mile wide right in the heart of the city. This is crossed by land vehicles only at its southern end, over a highway and by the S Bahn (the urban electrified railway operated by the Deutsche Bundesbahn). There are also water crossings in passenger craft operated by the Hamburger Hochbahn Aktiengesellschaft (HHA) who also operate the trams, buses and U Bahn. Of the total of all such urban passenger traffic, excepting the S Bahn, the U Bahn carries about 20 per cent, the trend being to replace surface tram routes by bus and underground routes.

Horse trams were introduced into Hamburg in 1866, and twenty-eight years later the first electric trams ran. In 1906 work began on an inner circular U Bahn line (opened in 1912) which was built partly on viaduct, partly in open cutting or at surface level, and about one quarter in tunnel. In the several extensions and new lines constructed since, this pattern has been continued; although latterly, construction in central areas has been mostly by the cut and cover method, box sections of reinforced concrete tunnel being prefabricated and lowered into trenches. This method lessened the period during which main thoroughfares along the line of route had to be closed to surface traffic.

In 1969 the U Bahn carried 179,200,000 passengers, but this figure will certainly be increased for 1970 when taking into account a new south–east line extension to Legienstrasse opened recently, and a new inner-city cross line that is still under construction, between Berliner Tor, Hauptbahnhof and Schlump. The S Bahn also, which was carrying about 360,000 passengers daily in 1967, is likely to show increased totals when its own linking line now under construction connects the present S Bahn terminus at Altona to Hauptbahnhof. The services of both systems were co-ordinated under a Joint Planning Board created in 1966, and given a common tariff and ticket system. Their lines are now numbered U1 to 3, and S1 to 6.

ADDITIONAL DETAIL

The U Bahn system: There was in 1968 51·7 miles (82·73 km) of route, all with standard gauge twin track, excepting the single-track branch from Volksdorf to Hansdorf.

Stations: The majority of stations are open, as most of the system is at surface level or on viaduct. Generally, those stations underground are at shallow level.

Current supply: Traction current at 750v DC is collected through a third rail.

Rolling stock: The trains consist of up to four DT2 type articulated twin cars, each twin carried on two end four-wheel trucks (bogies) and one central truck formed of two middle-coupled two-wheel trucks. They are driven by four 107 hp motors mounted on the end bogies. In the cab the driver controls speed and braking by foot pedals, the left for accelerating and the right for braking. Each train unit is 90 ft 3 in long overall and has a normal capacity of 260 passengers, including 82 seated. A new type, DT 3, consisting of three cars articulated, has been introduced on a large scale since 1968 to replace older stock.

Signalling: The U Bahn now operates under normal automatic block working with colour-light signals, but a new 'Train Guidance' system is at the time of writing under experimentation. Physically this comprises, basically, line conductors laid alongside each rail of the track that cross every 30 metres, and train apparatus enabling 'command' signals emanating from a fixed control point to be picked up by induction coils mounted on the cars. Through these conductor cables 'Control' sends commands for normal running and station stopping, in such a manner as to maintain safe braking distance between trains. These commands are 'programmed' or pre-determined to synchronize with train timetables. In effect this approaches full automatic working, the driver's job being to start the train, watch the line ahead and take control in emergency.

MUNICH

Munich, capital of Bavaria, has been endeavouring for years to cope with its growing street congestion, and at the same time to preserve the quality of its cultural centre. Like Frankfurt, it suffered heavily from wartime destruction, particularly at its centre, but has painstakingly restored its fine old buildings and rebuilt elsewhere, with characteristic German vigour, in the years following World War II.

Also, as with Frankfurt in particular and many other West German cities in general, prosperity has come to Munich. It has brought with it increased industry and commerce (employment in industry has grown threefold since 1939), and a widely expanded suburban area which in putting greater distances between work-people and their places of employment, has generated many more commuters. The former have shown an increased tendency in recent years to use their cars instead of crowded public transport, with the result that trams, buses and trolley-buses have suffered – not so much in patronage as in service delays due to their becoming ensnarled in traffic jams around busy road junctions.

The City Council's original plan for improving the flow of traffic included re-routing some of the tram services underground. But more comprehensive measures were called for after reappraisal of the city's present and future traffic requirements, and additionally for means to cope with the vast influx of tourists and others expected for the Olympic Games, due to be held in Munich in 1972. It was therefore decided that a conventional underground railway system was the best, and indeed the only means of mass transportation capable of effectively meeting the demands of a population presently at a million and a quarter, but likely to increase to a million and a half by 1990, with more than one million more in the immediate neighbourhood. Provision was also made for a special underground line to the site of the Olympic Games.

61

Actual construction of the first part of Munich's new U Bahn (underground) began in February 1965. This project, known as Line No 6, will extend from just beyond Kieferngartenstr. Station (above ground) in the north, through the city centre, and will terminate for the time being south of another underground line running east–west; that which is to be built for German Federal Railway trains. Geographically and functionally the first line of the Munich system resembles that of the Frankfurt system. Both are parts of future north–south transverse lines; both serve districts north of their respective centres warranting first consideration for rail rapid transit, and both have dual stations at the city centre for interchange with underground Federal Railway trains.

In the case of Munich the Federal Railway line (the S Bahn) will provide a much needed underground railway link between the central terminus, Hauptbahnhof, and the eastern terminus, Ostbahnhof. These stations handle a large amount of commuter traffic from the suburban lines that fan out from each station. Much of this traffic is obliged to use surface transport to cross the city, (there being no other direct link) so adding to the street congestion in this area. The S Bahn will be in tunnel partly built by the cut and cover method, and partly shield driven, and will convey nine-car electric trains, each capable of carrying 1,200 passengers.

The point at which it crosses the U Bahn Line 6 is at Marienplatz Station. This will a four-level underground complex, the upper sub-surface level for a passenger circulating area, the next two levels, one approximately above the other, for S Bahn trains going east and west, and the lowest level for U Bahn trains going north and south. The U Bahn line divides into two single tunnels just north of this point and comes together again just to the south. (This was to avoid undertunnelling of houses.) Thus there are two separate U Bahn stations at Marienplatz.

Otherwise the double-tracked line is in one rectangular concrete tunnel, 7·60 m wide, where it was constructed by the cut and cover method, and in circular concrete tunnel 5·53 m in diameter where it was shield driven. Incidentally, both types of tunnel are big enough internally to take ordinary suburban rolling stock, if ever this is necessary. They lie at a depth, measured from rail level, of approximately 46 to 52 ft (14 to 16 m) below surface level, except at Marienplatz station where they dip to 88 ft (27·0 m). The branch line to the Olympia site leaves Line No 6 just north of Munchner Freiheit Station. Construction work started on this short line in January 1968 and tunnelling is expected to take about two and a half years. Work on Line 6 is expected to take another two years (from 1969) but the first five lines of the U Bahn network proposed for Munich, totalling 35 km (including an extension of the Olympia branch line) are not expected to be completed until 1990.

The U Bahn trains will be of the conventional underground type, consisting of three two-car units, each unit having four 180 kw traction motors. A six-car train will thus total 2,600 to 3,000 hp and be capable of a maximum speed of 80 km/h and an average speed of 35 km/h. As each train will be able to carry about 900 passengers, the line, when operating at maximum train frequency, is likely to have a capacity of 40,000 passengers per hour in each direction.

ROTTERDAM

The status of Rotterdam rose to that of a world port in the late nineteenth century, when an artificial channel was constructed between it and the North Sea at Hook of Holland, enabling the largest merchant ships to reach Rotterdam on any state of the tide. Over subsequent years the dock area has spread along that channel until today the Europort–Rotterdam complex of docks and harbour facilities is the largest and busiest in the world.

Most recent reports quote the figure of 140 millions of tons of shipping handled in a year, as against 40 million tons pre-war. This emphasises the remarkable surge of dock construction that has occurred in the last twenty years, partly to handle new types of cargo and partly to handle a huge transhipment trade, including that of oil en route to Germany, transferred from ship to barge, and the two-way barge traffic between the port and the Rhine–Ruhr industrial region.

Associated with all the port activity has come also industrial activity, increased to such extent that the Europort–Rotterdam area is developing into a huge industrial centre where it is said that anything that can be made in a factory, is made. Industry, quotes the same source, accounts for two-thirds of the total of all output and employs two-thirds of the population. One third of a million people live on the south bank of the Maas, the river that flows through Rotterdam, so it follows that there is a considerable daily cross-river traffic (which is in fact about one-third of Rotterdam's working population). A few years ago the author remarked how much of this traffic was by bicycle – but this has probably been reduced by the advent of Rotterdam's Metro, which opened to traffic in February, 1968.

Prior to this, the river crossings consisted of a traffic tunnel and a traffic and railway bridge. The Metro now constitutes the third (under river) crossing and a fourth will be made when another road

1. *Above :* Barcelona: four-car train on Line 1; *centre :* Madrid: Sol ation; *below :* Madrid: two-car Metro unit.

12. *Above*: Stockholm: busy interchange on the T Bana system; *below*:
Stockholm: The T Bana in winter setting.

13. *Left :* Stockholm: centre street scene; *below :* Stockholm: bridge over Lake Malaren between Alvik and Kristineberg.

14. *Above*: Glasgow: underground car being raised into the maintenance shop; the underground routes do not reach the surface at any point; *below*: London: 1932 photograph of Piccadilly and Metropolitan trains together.

tunnel is completed. The Metro system's ultimate function when complete is to link developing suburban areas in the south with the city centre; and by a north-east spur, to serve also a new suburb of Rotterdam on Prins Alexander Polder, whose planned population is no less than 160,000.

The difficulties of building Rotterdam's underground-elevated Metro become apparent with the realization that Rotterdam for the most part lies beneath river level and is protected by strong river-bank dykes. Owing to the presence and pressure of water around the working sites, normal construction of the river approach tunnel segments *in situ* within excavated trenches was impracticable. Advantage was therefore taken of the presence of water in so far as it could be used as a transporting agency. The procedure adopted was to cut canals open to the river on each bank. The river approach tunnel segments, constructed with temporary bulkheads in building docks, were then floated into the canals and sunk in their appointed positions.

This was accomplished by building transporter rails on each bank of each canal, fitted with transverse beams. The segments were attached to these, and after being partially ballasted, were guided to their exact lowering positions. A somewhat similar procedure was adopted for the 'city' section where the tunnel segments, including parts of the 'Stadhuis' station consisting of platform sections and the hall above them, were built in a dock near the Metro site and floated into position. This semi-marine work was carried out within central areas of Rotterdam, which was thus presented with the spectacle of an excavated dock, with a small harbour in front for manoeuvering the segments of tunnel, in amongst the commercial traffic of the city.

The cross-river tunnel segments were constructed in a dock excavated on a near-by island in the river. They were subsequently attached to a girder bridge structure whose ends rested on pontoons. Rows of piles were driven into a previously excavated trench in the river bed and the segments lowered on to these. The union of segments one to the other was made watertight by fitting rubber jointing at the segment ends and bringing the segments together under hydrostatic pressure. The under-river section of tunnel lies at a depth which allows unimpeded passage for shipping. This section lies 6 ft 6 in (2·0 m) below the river bed, the tunnel then ascending under the city area until its top lies 11 ft 6 in (3·50 m) below street level.

South of the river the Metro, carried on viaduct, presents an aesthetically pleasing outline, far different from the elevated railways of early years. This is made possible by the wide use of pre-stressed concrete allowing the whole structure to be carried on single tapering piers. The track structure consists basically of longitudinal beams of concrete laid on to recessed transverse beams, which in turn rest on the piers. In tunnel sections the running rails, cushioned by semi-absorbent pads, are attached to longitudinal beams and cross sleepers dispensed with. Cost and labour was thus reduced by a reduction in the tunnel profile. There are four sub-surface stations on the city portion of the Metro and three at elevated level on the viaduct portion. Escalators are installed wherever necessary. Platforms 394 ft (120 m) long accommodate trains of four articulated twin-coach units, whose normal capacity is 1,160 passengers per four-unit train. In the first year of operation up to February 8, 1969, the Metro carried 33 million passengers.

ADDITIONAL DETAIL

Authority: Gemeentewerken Rotterdam.

Length of line operating in 1969: 3½ miles.

Rolling stock: twenty-seven coach units serving the line initially are being augmented to reach a total of forty-three. Each unit consists of two coach bodies resting on three trucks (bogies). On each side of each unit there are four double and two single doors. Capacity per unit is normally 290 passengers including 80 seated. Coach units are 95 ft 1½ in (29 m) long over buffers and 8 ft 3 in (2·67 m) wide.

Traction current: Collected by shoes acting on the underside of a third rail. Voltage, 750 DC.

Signalling: The present Automatic Train Control system with cab signalling, controlled from a main signalling station at Hilledijk stabling yards, is expected ultimately to be replaced by full Automatic Train Operation.

Line Capacity: Calculated at 35,000 passengers per hour in one direction during maximum train frequency (thirty trains per hour).

Electronic ticket scanning apparatus operates in conjunction with automatic turnstiles.

BUDAPEST

Budapest is one of the few remaining world major cities that has no heavy traffic congestion problem so far. Other cities that are building new Metro systems or expanding existing systems are doing so in many cases to alleviate chronic traffic congestion. Budapest is building its Metro partly to prevent future traffic congestion, but mainly because this growing city urgently needs a modern, high capacity underground railway system to relieve its public transport services, that are having to cope with an ever increasing flow of passenger traffic.

In Hungary, still predominently agricultural although rapidly industrializing, there is no other large town or city approaching the size and importance of its capital, and no other city ever likely to challenge Budapest's position as the country's cultural and industrial centre. Budapest's population of about two millions is expected to remain unchanged, or increase only slightly following a decentralization policy and falling birthrate; but even so, it will probably expand beyond its present 77 square miles (525 sq km) in future years by the inclusion of other districts in its metropolitan area.

There are two distinct parts today; the old Buda with its historical buildings and monuments occupying hilly ground on the Danube's west bank, and the newer commercial, industrial and residential Pest spreading over a large area of the flatter east bank, the twin cities being linked by eight bridges over the river. From the top of Gellert Hill, beneath the great memorial to the city's liberation from the Germans, the city view is extremely fine. Buda, dominated by its huge Baroque Royal Castle, is centrally a huddle of narrow, often hilly and winding streets. Pest is a geometric pattern of immensely wide streets and avenues, easily picked out at night by the lights of a dozen large stores and hundreds of small shops that do a steady business until a late hour.

This is the tourist's city, architecturally grand over a large central area, where articulated buses and coupled tramcars, crowded during most of the day and astonishingly crowded during rush hours, are all part of a bustling scene. For the average Budapest citizen these are at present the only means of city public transportation, representing a remarkably cheap and efficient means of travel, commuter or otherwise. There is a flat fare of one and a half forrints, with special low-price season tickets for pensioners and students, and free travel for children under six years.

Budapest already possesses a form of underground railway or tramway, but it is a short line carrying only a small proportion of the city's travelling public. Historically, however, it is notable as the first of its kind on the continent of Europe (being opened May 1896). It runs mostly at sub-surface level from Vorosmarty Square in the city centre, to the City Park about 3½ km to the north-east, and has eleven stations. The line is planned to be extended a short distance eastward from City Park, and to have its station platforms extended to take three-car trains. Until 1960 its trains comprised only one car, but a control trailer has since been added. The original cars have been reconstructed from time to time over seventy odd years and have been added to, to bring the present total to nineteen. The old railway's busiest time is during the summer, carrying families to the City Park's various amusements. The cars' driving current is taken from overhead wires at 525v DC.

Budapest's surface transport fleet comprises 1300 buses (many articulated with trailer), 2000 trams and 200 trolley-buses, and very frequent services are run. Nevertheless, overcrowding occurs daily on central street routes, principally for the reason that in the last few decades a variety of industry has located itself within a rough sixteen kilometers radius of the city centre, generating much passenger traffic in the city – especially from suburban railways, and between the main-line railway termini. This traffic must almost all cross the city by bus, tram or trolley-bus, and most of the overcrowding occurs along the Lenin Korut and continuing streets linking the East and West termini. To the large total of commuting passengers must be added the additional riders encouraged by the privileges mentioned earlier and by the low fare structure generally.

Were it not for lack of finance, Budapest would have had its first conventional Metro system operating in 1954. In 1949 the City Council and the Ministry of Transport ordered preliminary work to begin on an east–west line – a follow-up, in fact, of proposals for such a line that were submitted in the 1930s and negated by the subsequent preparations for war. One reason for the revived Metro momentum, springing from industrial development, was the big increase in tramway passengers from 380 millions in 1938 to 600 millions in 1949.

This proposed earlier line would have been largely a Russian conception, tunnel-driven with Russian shields and excavators, technically following in most respects the successful Russian Metro projects, and architecturally resembling them in above-ground station buildings, designed internally with ornamentation of the period.

Metro history, however, was again to be made in Budapest. Much of the actual tunnelling on the line of six and a half miles was completed when, in 1953, manpower and material had to be diverted to more essential work (housing and industrial building). The tunnels, partly at deep level excavated under compressed air conditions and iron-lined, and partly of 'cut and cover' construction, were abandoned except for periodic inspection and maintenance. They remained in this condition for ten years until 1963, when work on the Metro was resumed.

The Metro concept today is still marked by Russian influence. Much of the 'know-how' and all the escalators and rolling stock, for instance, are of Soviet origin. But in most other respects it is a Hungarian project. This includes the structure and design of deep-level stations and the bolt-less precast tunnel linings, as well as the general design and finish of the station approaches, etc. The stations are bright and colourful, but simplicity and functionalism are aimed at rather than ornamentation, as there is no doubt that pressure of traffic will build up in these stations as the Metro develops from this initial ten-kilometre line.

The first five and a half kilometres of the new line from Feher Road in the east to Deak Square is expected to be opened in 1970*.

* The Metro as far as Deak Square was in fact opened on April 2, 1970.

The remaining four and a half kilometres, extending into Buda, is expected to be completed by 1972. The line's car depot with twelve eight-car stabling tracks is near the open Feher Road terminus. It will have all facilities for car maintenance, and up to major overhaul, involving complete stripping down normally after 50,000 km of running. The depot features the customary 'blow-out' and lifting shops, situated on a large site, indicative that the depot may be subject for expansion as the Metro network develops.

This locality around Feher Road on Budapest's outskirts, flat and with much open ground, is understood to be planned for development when the Metro is operative. As it is, there is the well-known 'Icarus' works near-by, with 6,000 employees producing motor-buses, along with a variety of heavy equipment, and near-by also is the Zugle residential district. From here the Metro descends into cut and cover tunnel to the first underground station at People's Stadium, then dips sharply to deep level to Baross Square, in front of the principal main line East station. The Metro station here is being overbuilt with a sunken precinct open to the sky, reached by radial passages from four or five streets. A bank of three escalators leads below-ground to what will probably be the busiest station on the Metro.

The next three stations are Blaha Lujza Square, Astoria and Deak Square, the latter joining with Engels Square to form almost a small park at the city's hub, where motor-coach parties assemble. The Metro, in twin iron tube, is at its deepest level (120 ft or 36·5 m) at this point. The line then continues to Kossuth Square, in front of the Parliament Building, then crosses beneath the Danube just south of the attractive Margaret Island to Batthyany Square in Buda (for suburban railway interchange), then west to Moskva Square before swinging south, to terminate in front of the South railway station terminus.

Most of the Metro stations lie beneath large, well-lit concourses, themselves located beneath the squares that give the stations their names. The only evidence of the Metro above-ground will be the many entrances at street intersections. The central control building for the east–west line is at Deak Square, where staff will be in direct touch with train operators by radio telephone. The trains, built at the Mitishinsky Works near Moscow, will be in yellow livery. Unlike London Transport stock their driving control incorporates no 'dead man's handle'. To safeguard the train there are two men in the cab; a motorman, and a guard whose duties include operation of the four sets of double doors (each side), pneumatically actuated.

Although the track is standard gauge (1·435 m) the car bodies are of normal Russian width, 8 ft 10 in (2·67 m) their two bogie or truck frames being modified to bring the running wheels closer together. Traction current at 825v DC is fed through a third rail by top contact. Fail-safe apparatus co-ordinates the two braking systems, rheostatic and pneumatic, and train-stop mechanism as featured on most lines is incorporated. Passenger capacity is 170, including 42 seated, but 220 passengers per car can be carried at peak periods.

In view of the bogie modifications, more than usually intensive test running is being undertaken (1968–9) on experimental track 2·5 km long, forming part of the east-west line at Feher Road. It is ideally suited for the purpose as one half is on the level in the open and the other half is on a slight incline, almost all below-ground, leading to the tunnels and continuing along them. An experimental train of four cars is being run back and forth along this section continuously, until it reaches a total running distance of 100,000 km. The author, privileged to ride on this train, paid special attention to the motion in view of the train's bogie modification, but experienced only smooth acceleration to about 70 km/hr and no rolling or nosing. The track in the open is secured to concrete sleepers (ties) by clips and insulated screws, and is set in concrete in the tunnels and continuously welded. The running rails weigh 48·3 kg/m. Minimum radius of curvature on the line is 400 metres.

Stations, averaging 1110 yards distant from each other, generally have central island platforms, all of which are 396 ft (120 m) long, sufficient to accommodate trains of three two-car units, although initially only four-car trains will operate. Service speeds are expected to average 33 km/hr, with a maximum acceleration and deceleration speed both of 1·5 m/sec. The line's conventional signalling system using colour-light signals will permit a service density of forty trains per hour, with thirty-second station stops. It will be manufactured under licence from the Swiss Indentra System. Maximum line capacity is expected to be 48,000 passengers per hour.

The banks of three escalators being installed at each below-ground station will be inclined at 30° from the horizontal. All are reversible and will operate at an average speed of 0·9 m/sec. The highest planned vertical rise on escalators on the first line is 96 ft (29·6 m).

Construction of a second Metro line for Budapest is planned to start in 1969. This line, running south-east, will be entirely on the Danube's east side and will be about nine miles long. It will start in the north at Istvan Square and run south via Marx Square to Deak Square (for interchange with the first line) and Felazabadulus Square to Calvin Square, where it will turn south-east under Ulloi Street to terminate at Hafar Street. A third Metro, a branch off the north–south line, is also scheduled under a twenty-year road–rail development plan. The first line is expected to carry 300,000 passengers daily, equalling 8 per cent of the public passenger transport total: the second line will raise the Metro percentage to about twenty, and the third to about thirty per cent. Thus even when all the lines are operative the majority of passenger journeys will still be by public surface transport.

MILAN

Milan, with nearly 1,750,000 people in its metropolitan area, has been described as the largest industrial centre in northern Italy; but the city has also increased its commercial activity (as witness the many recent office and warehouse buildings) so that today it could more accurately be called Italy's leading financial, industrial and commercial city, second to Rome in size. Milan's geographical position, on the Lombard Plain, close to Alpine passes leading to northern Europe, and therefore on principal trade routes, and subsequently trunk railways and motorways, has ensured its accessibility and importance throughout history. The past, however, conferred on the old city centre a pattern of narrow streets that inevitably become congested with modern surface traffic – a condition that is aggravated by the presence of the otherwise excellent tram and trolley-bus services.

It was decided some years ago, when traffic increases became an accepted fact, that Milan's public transport needs (and indirectly its social and economic future) could best be served by a modern rapid transit or underground railway system that would provide adequate mass transportation, and progressively replace the trams, starting with the congested inner city routes.

The first move towards this objective was made in 1953, when a general project was put forward for governmental approval by the Municipality. By stages the point was reached when, in 1957, approval for an experimental section of underground railway was obtained, and in 1958 the go-ahead was received for the first complete transverse city line (Line 1). When eventually this line was completed and opened in November 1964, it was one of the first entirely new 'underground' projects in Europe built in post-war years.

New procedures were followed in its tunnel construction (by the

74

cut and cover method) to minimize the danger of settlement where it passed close to historic buildings (including the famous Cathedral), and the period during which traffic in the busy streets was impeded. The vertical tunnel walls were built in reinforced concrete by a process which largely dispensed with vibrating machinery. Basically this consisted of digging narrow trenches progressively deeper as a form of clay slurry known as Bentonite was poured in, the Bentonite serving to retain the trench walls. Liquid cement was then pumped in at the bottom until it reached the required height of the walls, and in so doing it displaced the bentonite. When these side walls had set, soil was removed between them to roof-bearing level, the tunnel roof cast *in situ* and the streets opened to traffic on the temporary filled in surface. Thus it was possible to carry on with the actual tunnel excavation whilst traffic ran almost normally overhead.

The present line runs entirely in double-track rectangular tunnel, generally 10 ft (3 m) below surface level, from Marelli in the north down to Piazza Duomo (Cathedral Square), where it turns westward to Lotto station, with a short branch south-west to Gambara station. Today the Metro lines total 14·2 km, but work is proceeding on Line 2 which will run eastward from near the Central Station of the Italian State Railways to link with the Ferrovie Celeri dell'Adda (a tramway upgraded to rapid transit standards) reaching out about 13½ miles (22 km) to Gorgonzola. When the former line is completed the entire length, including the Gorgonzola rail route, will be known as Metro Line 2.* Thereafter the proposal is to build two more Metro lines to connect south Milan with the centre.

The Metro was designed from the beginning to conserve manpower, in respect to its operation. One man operates a train and one man controls the entry barriers at each station. He is enabled to exercise supervision by a system of closed-circuit television which permits him a view, through a monitoring set, of the platforms as well as the entrances. Train movement over the whole line is under central traffic control. Communication from train to traffic control is effected through plug-in telephone sets from frequent points

* Line 2 is now (1970) partly operative.

along the tunnels, and there is telephonic communication between control and every station.

ADDITIONAL DETAIL

Authority: Azienda Transporti Municipali, which also operates the trams, buses and trolley-buses.

Stations: There are twenty-four at present, with ticket halls between street level and station level, and served by escalators. passenger barriers across station entrances are actuated by the insertion of tickets, which are scanned electronically. There is a basic fare of 100 lire on the Metro. Station platforms 106 metres long accommodate trains of six cars.

Rolling stock: The cars each have a single control cab and are driven by four 120 hp motors. Braking is by a rheostatic and pneumatic system. Rubber inserts between hub and steel rim of the cars' running wheels assist in smooth running. Passenger capacity on the two types of car is 200 and 213 passengers, with seats for 27 and 30 respectively.

Track gauge: 4 ft 8½ in (1·435 m).

Signalling and traction current: The automatic block system with colour-light signals also repeats these signals in the driver's cab, indicating a range of four permissive speeds, and stop. Traction current at 750v DC is collected by the trains through shoes contacting a third outer rail, and returns through a fourth rail. Line 2 is powered at 1,500v through an overhead system.

The yearly total of passengers carried by the Metro increased from approximately 36 millions in 1965 to approximately 56 millions in 1968. In the latter year approximately 576 million passengers were carried on the whole system of Metro, tram, bus and trolley-bus lines.

ROME

In recent years Italy has vastly improved its land communications, both to encourage tourists and to link the industrial north with new industrial centres in the south. Rome lies equidistant between the two, on motorway and rail routes, famous first as a centre of ancient culture and latterly enhanced in importance by the growth of industry, and by improved travel facilities that heighten its attraction as a tourist centre. Whatever the intrinsic reasons for the city's remarkable growth, this is confirmed by population figures that in the last century have roughly doubled every thirty years, and today approach the two and a half million mark.

There is, too, a consistent increase in the number of passengers carried by the Rome Metro in recent years, although this is a very small percentage of all those carried by Rome's surface public transport. The Metro was not designed to be a general carrier originally, being but a short line (6·8 miles or 11·03 km) built for a specific purpose – to provide mass rail transportation between the city and the World Exhibition planned to be held on a site south of the city in 1942. The project was partly completed, tunnelling being almost finished, when work was suspended and remained so until after the war. It was resumed to alleviate unemployment, and in prospect of the Exhibition site being developed (although not to its original concept, as this did not find favour in post-war circumstances). The Metro's passenger total for the year 1964 was seventeen millions, which increased in 1968 to nearly twenty millions.

Today, most Metro trains run between the Termini station and Laurentia, but some continue via a connection at Magliana to Lido di Ostia on the coast. (The private Company operating the Metro also operates the Rome-Lido line.) The Metro for 3·7 miles of its length runs underground, being in tunnel from Termini, beneath the main-line terminus, to beyond S Paolo station, and again under the Exhibition site in the south.

The more populous areas west and east of central Rome, however, are also to have a Metro line. Designated Line A, this new line is under construction and will add some nine miles to the system. It will run from Risorgimento eastward to Osteria del Curato, connecting with the existing line at Termini. These two lines will in future form the nucleus of a much larger Metro network, if plans now existent come to fruition.

ADDITIONAL DETAIL

Stations : There are eleven stations at present, with ticket offices on the surface in some cases, and others underground.

Rolling stock : On the Termini-Lido service trains are made up of six cars, and on the Termini-Laurentia service, three. They are 63 ft 8 in (19·4 m) long over the buffers and have a capacity of 240 passengers, including 48 seated.

Signalling : At present the automatic block system, with track circuiting, is used, but cab-signalling may be adopted, permitting a frequency of forty trains per hour.

Traction current : This is taken through four sub-stations from the public supply and fed to the trains by an overhead system.

Tunnels and Track : The tunnels are double-track, elliptical, 26 ft 4 in wide (8·02 m) and 18 ft 1 in (5·5 m) high from rail level. The track is standard (4 ft 8½ in) gauge, laid with flat bottomed rail.

Authority : Societa della Tramvie e Ferrovia Eletriche di Rome (STEFER).

In 1967 the whole Rome Public Transport system (buses, trams and trolley buses) carried 775,750,333 passengers.

OSLO

There are certain similarities between Oslo and Stockholm that will be noticed when reading the section describing the latter city. Both Nordic capitals have felt it necessary to provide improved public transport in recent years, partly to replace that no longer adequate, and partly to ease congestion caused by increasing car ownership and use, the last named reflecting the rising economy of both countries. Physically, too, Oslo resembles Stockholm in so far as it is much the largest city in the country and has developed considerably its trade and industry in the last two decades.

Like Stockholm, Oslo has turned to rail rapid transit as the best means of meeting present and future demands for both social and commuter travel. Its present population of about half a million is expected to increase to 540,000 by 1980 and to one million by the year 2000. After World War II, first considerations were for a considerable housing programme, and as external and internal trade developed there followed a gradual expansion of industrial areas, mainly on the east and south-east side of the city. The system of tramways or light railways, and buses, could neither be adapted to adequately increase capacity, or their routes be expanded sufficiently on the surface areas available. This latter is explained by the fact that the old part of Oslo, at the head of a fiord, lies sheltered in a bowl-shaped declivity, with land rising to more than 1,800 ft around it. The centre is thus a concentration point for traffic movement and inevitably attracts more vehicles than there is road space to permit an unimpeded flow.

The solution lay in a form of tunnel-borne mass transportation, and despite the considerable difficulties of construction, tunnel lines have been built, partly through hilly ridges, and an underground railway system now links the east and south-east suburbs of Oslo to the city centre. It does not yet link up with the western tramway system as was projected, although as far back as 1928 this

link was shortened when the common line of three western tramway systems was extended cityward in a long double-tracked
tunnel one and a quarter miles long, from Majorstuan to National
Theatre. Although there may be an underground link eventually,
no specified year has yet been announced for a start on the work.

The difficulties in constructing the inner portion of the three-
line eastern network may be partially illustrated by the fact that the
lowest point in the main (common) tunnel under Akerselva is 26
ft (8·0 m) below sea level, whilst the highest points on the branch
lines are from 510 ft (156 m) to 600 ft (183 m) above sea level.
Furthermore, the tunnels in rock amount to approximately five and
a half miles and those in clay, in reinforced concrete, to approximately one mile. The system now operative consists of the new
lines from Jerbanetorget to Halse and Helsfyr and that to Skull-
erud, which was an old tramline reconstructed to rapid transit
standards.

The total operative route length is 15·7 miles (26·8 km), but a
fourth line under construction and due for partial opening in 1970,
the Furuset, will bring the total to 21·75 miles (35·0 km). Additionally, a 2·5 km extension from Grorud in the north-east to a newly
planned suburb, and an eastward extension of the main line to a new
city terminus (Sholtsparken Station) are respectively being constructed or planned.

ADDITIONAL DETAIL

Authority: A/S Oslo Sporveier.

Number of operative stations (1968) thirty-three. Five inner stations
are underground. All stations have side platforms 110 m long to
take 6-car trains. They are spaced 0·57 miles (815 m) apart, on
average.

Rolling Stock: There are 105 cars, each powered by four 98 kw
motors, The braking system incorporated dynamic, air and hand-
braking. The capacity of the cars is 97 passengers standing and
63 seated. A car's overall length is 55 ft 9 in (17·0 m).

Track gauge: 4 ft 8½ in (1·435 m).

Signalling and Power: Coded signal impulses from the track, picked

5. *Above:* London: Victoria Line, Coburg Street control room, showing illuminated diagrams, desks for Traffic Controller and Signal Regulator, and television monitor screens; *below:* London: model of the Oxford Circus complex of tunnels. The Central Line runs left to right in the background; the Bakerloo and Victoria lines are in the foreground.

16. *Above:* London: exterior of automatically driven Victoria Line train. The coils which pick up signal and driving commands can be seen in front of the wheels; *right:* London: automatic fare collection.

17. *Above:* Moscow: interior of Metro carriage; *below:* Moscow: Arbat station vestibule.

18. *Right:* Moscow: Electroza-vodskaya station vestibule; *below:* Leningrad: typical station.

up by train apparatus and converted to visual signals in the driver's cab, permit speeds through the range 70–50–30–15 km/h. A traffic control centre at Töyen controls the train services. There is both radio and telephonic apparatus on the train. Power for traction is collected through a third rail at 750v DC, obtained through rectifiers from the city's 5000v AC supply.

LISBON

Portugal's industrial growth rate in recent years, and hopes for expansion under the third Development Plan, must inevitably contribute to the importance of Lisbon, capital city and the only seaport of any excellence along the Portugese coast. Lisbon's prestige is likely also to be progressively enhanced following the opening three years ago of the huge River Tagus suspension bridge, the longest in Europe, whose northern end is about three kilometres west of the city centre. Further, the bridge may well give impetus to the project for expanding Lisbon's Metro system, at present comprising one main and one branch line, into a multi-line network encircling the city and extending east and west to serve built-up areas along the coast.

Historically, Lisbon's urban transport has progressed from horse drawn coaches or buses, through mule-drawn trams to electric tramways introduced in 1901 by the Compania Carris de Ferro de Lisboa; and finally to the opening in December 1959 of the city's first Metro routes beneath the famous Avenida Liberdade to Rotunda, and then continuing north-west to Sete Rios and north to Entre-Campos. The main line was extended south-east into the commercial centre in 1963 and then north-east along the Avenida Almirante to Anjos in 1966. It is operated by the Metropolitano de Lisboa, SARL, a private Company, the largest share of whose capital belongs to the Municipality.

Further Metro extensions proposed will first be along the main radial routes, where existing trams will be replaced by buses. The final Metro network of lines proposed is expected to carry one-third of Lisbon's public passenger traffic, with surface public transport carrying the remaining two-thirds. Metro patronage has increased with each successive Metro extension, the figures being 26 millions for the year 1966, 33 millions in 1967 and nearly 37 millions in 1968.

82

The trains comprise just two motored coaches at present, carrying about 400 passengers, but the stations were constructed with a provision for possible lengthening of their platforms from the present 131 ft (40 m) to 230 ft (70 m) in order to accommodate four-coach trains. The whole system is underground, but only Parque station, built at deep level, has escalators. The tunnels were built mainly by the cut and cover method and have arched roofs over vertical walls. The track is of standard gauge, laid with welded rail. Rolling stock at present consists of thirty-eight coaches built by Linke-Hofmann-Busche, and Sorefame (Portugal). They are driven by 122 hp motors mounted on each axle and are 54 ft 1½ in (16·5 m) long overall.

Signalling incorporates the automatic block system with colour light signals. Traction current is taken principally from the national grid, through one main sub-station and two traction sub-stations, and is collected through a third rail system at 750v DC The Metro workshops are near Sete Rios, the present terminus of Line No 1 bis.

BARCELONA

Barcelona, described some time ago as the most prosperous and most populous seaport in Spain, has consolidated its position in recent years and appears certain to maintain it into the foreseeable future. Since ancient times the city has attracted trade to itself through its natural harbour, and with each successive harbour improvement this trade has increased. Industry established itself here soon after the coming of the steam railway in the early nineteenth century, and the combined effects of sea trade, the railway advent and industry progressively extended the building limits outward from the nucleus of narrow streets by the harbour which once constituted the entire city.

Between 1897 and 1922 several suburbs were absorbed into Barcelona and now the city fills practically the whole natural hollow contained by the hills around the city. Barcelona's recent growth is illustrated by population figures that in 1950 were little more than one and a quarter millions and today approach the two million mark. To cope with the present and anticipated demands for mass public transportation in areas so far not served by the existing Metro, the authorities have planned a massive expansion of its present three-line system, totalling about thirteen route miles, to one of seven lines totalling nearly forty-nine route miles.

The present system, operated by the municipally controlled FC Metropolitano de Barcelona SA, comprises three lines that were until 1961 independently operated. The longest, once named the Transversal and now Line 1, is a broad gauge line of 5 ft 6 in (1·674 m), originally designed as a cross-city connecting line for the National Railways and therefore conforming to the latter's track gauge. The city's first Metro line, originally the Gran Metropolitano and now Line 3, was opened in 1924. The second, the Transversal, was opened in 1926 and the third, previously the Sagrera–Horta line and now Line 2, was opened in 1959. The proposed Metro expansion is developing as extensions of Lines 2 and 3,

84

which are of standard 4 ft 8½ in (1·435 m) gauge, and as new lines of the same gauge.

Line 3 runs north–south from Lesseps to Cataluna in the city centre, where it branches south to Correos on the harbour front, and south–west to Liceo. This station, incidentally, is on the Ramblas, the street famous as a promenade, and for its cafés and flower-stalls, etc. The Liceo line is now being extended, under Phase 1 of the expansion scheme, towards the harbour, where it will turn and follow the Calle del Marques del Duero to Plaza de Espano, subsequently describing a half-circle to rejoin the present line near Lesseps, and so form a circular route serving the west part of the city.

The Horta line (No 2) is to be extended south-west from Sagrera to cross the mid-city and end at an interchange station on the new Line 3 extension. A second circular line, taking in the Correos No 3 branch line, is to be built to serve the eastern coastal district and the east part of the city generally. Additionally, two lines are planned to traverse the city from north to south and east to west respectively.

The new underground system will be known as the Red de Metro de Barcelona (Metro System of Barcelona). Its new construction work is scheduled to be undertaken over a period of eleven years from 1967. At present (1969) there are thirty-four operative Metro stations. The whole expansion scheme envisages a total of 108 stations, and would provide Barcelona with an underground Metro system affording coverage as comprehensive as that of any similar city in the world.

In addition to the municipally operated Metro system, there is also the independently operated Sarria Line. This originally formed part of the surface system of the Cataluna Railway Company, running from Cataluna to Tarrasa, eighteen miles northwest of Barcelona. The urban part of the line was subsequently rebuilt as a tunnel line and therefore forms part of Barcelona's underground railway system. The underground part runs as far as Sarria and there is a branch leaving the Sarria line at Gracia, to Avenida Tibidabo at the foot of the hills. The Sarria line and branch together total about four and a half miles. Its cars derive their driving current from an over-head wire system at 1300v DC.

Tunnels: The whole of the Metro system is in tunnel with the exception of a short stretch of line at the west end of Line 1. The tunnels are double tracked, those of Line 1 being 26 ft 3 in (8 m) wide, and those of Lines 2 and 3 being 23 ft (7 m) wide.

Traction current: There is a third rail system on Line 1 carrying traction current at 1500v DC. Lines 2 and 3 both derive their current from an overhead wire system at 1200v DC.

Rolling stock: Because of their independent histories the three lines together carry a variety of types of car. Those on Line 1 are of four classes, 100, 200, 300 and 400, the first being built in 1926. The stock consists of motor-cars and trailers. Line 2 operates with all-motored cars built in 1959. Line 3 operates with four classes of car, the earliest built in 1923, the stock being made up of both motor-cars and trailers.

Stations: The Metro stations vary in lay-out due to the same reason as quoted above. Two stations on Line 1 are common with the Spanish National Railways. Central platforms and side platforms both feature in the total of Metro stations.

Signalling: The basic system is by automatic block with AC track circuiting and colour-light signals. Since 1964, however, Line 2. which operates with only one type of car, has been equipped with a form of automatic signalling using sensitive photo-electric cells and metallic screens. The latter are fixed along the track at points where deceleration or braking are desired. Relays in the driving-motor control system are actuated, cutting off current to the motors, whenever the screens cut off light to the photo-electric cells. The same action is applied to the braking control when down-gradients or signals call for braking; all being subject to safeguards in that the driver can operate manual control when necessary.

In 1968 the Metro system carried approximately 203½ million passengers, a figure that will normally increase as further sections of extended lines are opened. Barcelona also has an extensive system of surface transport comprising trams, buses and trolley-buses.

MADRID

Madrid's rise in population from about three quarters of a million people a century ago to nearly three millions today may be attributed to a number of factors. That it is a university city and the centre of education for northern Spain is one. The many light industries that established themselves there in the early twentieth century is another, in so far as they contributed to Madrid's development – and to these two can be added Madrid's central position in Spain on railway and road routes; and a natural migration of people from country districts to seek higher living standards in town.

Madrid's rapid physical expansion in the years preceding the civil war, 1936–9, was due in part to the enlargement of its Metro system, especially to the north and east of the centre where suburban living was facilitated by the Metro's extension to these areas. In another direction, development has been equally rapid in recent years, across the river Manzanares and to the south-west as far as the suburb of Carabanchel. This suburb has for many years been connected to Madrid by the Ferrocarril Suburbano, a railway operated by the Metro, but it passes through less development-prone areas than the new Metro line which now links Carabanchel directly to Madrid's main entertainment and shopping district. This line, and other proposed new Metro lines, are part of an ambitious plan to provide Madrid with a Metro system capable of carrying by far the largest proportion of the city's travelling public by the 1980s, when Madrid's population is expected to reach four millions.

Just now (1969) it carries more than all the tram, bus and trolley-bus services put together. This is remarkable, considering the many services run by the latter; but for some years the Madrid Metro carried more passengers per kilometre of route than any other similar system in the world. (Tokyo and one or two other systems now slightly exceed the Madrid figures.) The people of Madrid use their Metro extensively, partly because until quite recently the network of lines was confined largely to densely built-up areas, and partly because, like Paris, its stations are situated close together and are easily accessible from street entrances by stairs. The average journey length of 3·5 km is relatively short – about half that on London's system.

The history of the Metro is one of consistent expansion since the first three and a half kilometres of line was opened in 1919, from Cuatro Caminos south to Puerta del Sol, the traffic hub of Madrid: and at the time of writing, another stretch of line, extending the Carabanchel line across Madrid for 2·8 miles (4·6 km) to link first with the De Leon line and then with the Cuidad Lineal line at Ventas, is under construction. Enlargement of stations, plus acquisition of new rolling stock and improvement of track, sub-stations, and signalling, have all been part of the Metro development plan mentioned earlier.

ADDITIONAL DETAIL

Authority: Compania del Metropolitano de Madrid.

Number of lines: Six.

Length of route: (1969) 25 miles (40·07 km).

Stations: The largest and most important of the system's seventy-one stations (1968) is Sol, where three lines cross, and escalators are installed. Otherwise the stations, in single elliptical tunnel, are at shallow level. A programme of lengthening platforms generally to take six-car trains is under way.

Tunnels and track: Generally the running tunnels are rectangular, built by the cut and cover method, and by the 'gallery' method, containing double tracks of 4 ft 8⅞ in (1·445 m) gauge.

Rolling stock: the older stock includes both motored cars and trailer cars, but the latest stock, built by Genemesa and General Electrica Española, are all motored cars, driven by four 100 hp motors. Trains of this design are going into service as delivered for the Carabanchel line and for replacement of cars on other lines. They form trains of all-motored cars. The overall car length is 46 ft 11 in (14·3 m).

9. *Above:* Kiev: vestibule of University station; *below:* Kiev: Kresh-
atik station.

20. *Above:* Tbilisi: surface vestibule, Rustaveli Square station; *below*
Tbilisi: Didube station.

21. *Above:* Tbilisi: Vokzalnaya Station platform; *below:* Baku: Gyandi-zhlik station platform.

22. *Above*: Montreal: train approaching Rosemount station; *below*: Mo
treal: rolling stock.

Traction current: The 15,000v public supply is transformed in seven sub-stations to 600v DC.

Signalling: The automatic block system with AC track circuiting and colour-light signalling is used. A system of automatic train control (ATC) is in operation on Line 5.

Workshops: the main ones are at Cuatro Caminos and there are smaller workshops at three other points.

In 1967 the Madrid Municipal Transport system carried 435 million passengers on its trams, buses and trolley-buses. The Metro in that year carried approximately 461½ million passengers. There is also a section of the Spanish National Railways running underground from north to south Madrid, separate from the Metro, but providing interchange with it at three points.

STOCKHOLM

If a capital city's public transport tends to reflect some of the country's characteristics, the Stockholm T Bana (Underground) might be said to have this tendency. Sweden, about three times the size of England and Wales but much more thinly populated, is relatively one of the richest countries in the world. It has a population of approximately eight millions, one sixth of whom live in Greater Stockholm. In the last twenty-five years the country has moved away from exporting its raw materials (timber, ores, etc.) to creating manufactures from them and exporting these. To have maintained and increased the standard of living during this period implies an increased application of technology to manufacturing processes, and this has occurred in Sweden. It is a forward-looking country, and its capital city's T Bana system reflects this characteristic.

Greater Stockholm has more than doubled its population since 1920. The nucleus of the city is the Old Town, where most of its historical buildings lie, but much of its suburban area has been re-built in comparatively recent years and lofty white apartment houses are now a feature. Overall, Stockholm gives the impression of a light, airy city, clean and rather modernistic. About half the total area within the boundaries of Greater Stockholm is open water, a feature which contributes to the brightness of the city, but added difficulties to the expansion of its metro system. This latter is one of the youngest in Europe and, as might be expected, is very modern-looking, in both its station design and decoration, and in its trains.

Its history is one of rapid development to cope with a correspondingly rapid increase in car ownership and the threat of too many cars congesting the city streets. The forerunner of the T Bana was an underground or tunnel line opened in 1933 between Slussen and Ringvagen on the built-up South Island of Stockholm.

It was used only for tramcars for seventeen years and so cannot be called the beginning of the T Bana. This came into being in October 1950, after the tunnel had been adapted for underground trains, which ran to the suburban terminal.

A year later a south-western branch line was opened between Gullmarsplan and Stureby, and this was followed a year later by the first line built specifically for conventional underground trains which ran from the north-west suburb of Vallingby to Hotorget in Central Stockholm. It was not until 1957 that the T Bana's northern and southern halves were linked together. This was accomplished by the building of a connecting line from Central Stockholm to South Island, which went by way of a tunnel under the Norrstrom (northern river channel) and by bridge over the Soderstrom.

The South Island and the southern mainland are further separated in the south-west by the fairly wide waters of the Liljeholmsviken, which had to be crossed when the T Bana was extended south-west towards its present two termini, Varberg and Fruangen. A crossing was made by way of a concrete underwater box tunnel which was constructed in dry dock, floated into position and sunk 24 ft (7·0 m) to the dredged bed of the channel. It is one of the early examples of crossing waterways by this method (mentioned in the main text) and illustrates just one of the complexities faced by the engineers when projecting the T Bana under and over water channels, across valleys and through ground that varies from rock to water bearing gravel and sand.

The T Bana is not only interesting in the variety of its terrain, and the seasonal changes that transform the latter from summer greenery to heavily snow-laden countryside. It also has a technical distinction, in so far as it was the first underground railway to adopt cab signalling. Coloured signal aspects in the cab are actuated by impulses picked up by induction from the running rails, through receiver coils. The motorman thus has a continuous indication of his permissible speeds, ranging from 70 km/h to 15 km/h.

The Stockholm system had in 1969 reached out to Hasselby Strand in the north-west, to Farsta in the south, to Ropsten in the north-east and to Varberg in the south-west, besides other branches. The farthest from Central Stockholm (Hasselby Strand) is about 9½ miles (15 km) distant, but extensions planned for the next seven years are likely to increase that distance, in the south-west, to about twenty km.

ADDITIONAL DETAIL

Authority: AB Storstockholms Lokaltrafik.

Length of route: 39·3 miles (63·2 km), of which 16·3 miles (22·5km) is in tunnel.

System Network: Two transverse lines with branches, all double tracked.

Track gauge: 4 ft 8½ in (1·435 m).

Number of stations: sixty-eight, spaced approximately half a mile apart on average,

Station lay-out: Generally the underground stations are approached by stairs, (or escalators if the depth warrants them) from sub-surface ticket halls. Decoration of station walls displays artistry in permanent form of mosaic, in some instances. Advertising by poster is limited.

Passengers carried in year 1968: Approximately 180 millions.

Number of cars per train: eight. Number of trains per hour, maximum forty-eight.

Passenger capacity per car: 140 (50 seats).

Traction current and collection method: 650v DC by third rail.

Rolling stock: In green livery. The cars are 57 ft 1 in overall length and have three double-leaf automatically operated doors per side. The latest cars have rubber suspension.

Signalling: The cab-signalling system obviates fixed lineside signals except at junctions.

Tunnel Detail and Line Contour: The tunnels are mainly rectangular in section, 24 ft 11 in (7·6 m) wide. The maximum tunnel depth is approximagely 43 ft (15 m) below surface level, but the lines' contours vary considerably and at one point the line is 156 ft (47·6 m) above datum level.

GLASGOW

Nearly two million people live in the Greater Glasgow area, by far the largest of all Scottish cities. Glasgow's period of greatest prosperity was between 1870 and 1920, when industrial production was at its height and ships built on the Clyde carried its goods all over the world. The city is now undergoing re-planning, old factories and dwellings being replaced on a scale said to be equal to any in Europe. Much attention has been given to easier communication, with emphasis on a most extensive urban motorway programme; but public transportation enters importantly into the overall scheme for Glasgow.

The city's 4-ft-gauge tube railway is one of the oldest in the world. It began carrying passengers in 1896, and now after more than seventy years' operation it still has a fairly consistent patronage, although a reduction in passenger totals from their peak reached about fifteen years ago must be viewed against the background of improved public surface transport operated by the Corporation (who also run the Underground) and increased private motoring.

Glasgow's Underground or Subway is unique in that its twin tunnels which follow a circular route around the inner city, have no physical connection one with the other, and the whole 6·6 mile line is underground. Its cars, which comprise an equal number of motored cars and trailers, are stabled on the running tracks and when requiring maintenance or repair are hoisted bodily up into the maintenance shop on the surface. The tunnels, in some places only a few feet from the surface, descend to deep level (maximum 115 ft) resulting in some stiff gradients, a feature which reflects the line's original cable railway design, when its trains moved or stopped as a clamp affixed to the leading car was tightened or loosened on the haulage cable. The railway was converted to electric traction in 1936.

LONDON

The two most important events in the later history of London's Underground system took place on January 10, 1963 and March 7, 1969 respectively. The first was the centenary celebration of the world's first passenger-carrying underground railway, which commemorated the opening in 1863 of that four-mile-long steam-operated line between Farringdon Street and Paddington, London. The second was the official opening of the Victoria line ten and a half miles long, by the Queen, who thus inaugurated the first completely new underground railway to be built in London for over sixty-two years. (The last such railway was the Great Northern, Piccadilly and Brompton line, opened between Hammersmith and Finsbury Park on December 15, 1906.) The London Transport system of railways, with $254\frac{1}{2}$ miles of route, is the largest in the world. Together with London Transport's buses and coaches it serves an area which, including Greater London's 620 square miles, contains a population of more than ten millions.

The history of London's Underground contains a wealth of technical and general information, beside a good deal of interesting anecdote: but in a book of this description the author can do little more than compress its century of history into a chronological record and a few deserving highlights. The most important of these is undoubtedly the adoption of electricity as motive power, which made possible and practical the transportation of humans in large numbers through 'Tube' tunnels driven deep below ground.

In such tunnels, with almost complete reliance on artificial ventilation, steam locomotion would have been out of the question (as would later have been other forms of motive power deriving from combustible fuel). Thus all the newly invented tunnelling techniques, starting with Peter Barlow's shield used on the second Thames Tunnel and continuing with the famous Greathead shield which cut most of London's Tube tunnels, would never have been

94

successfully exploited in the field of urban transportation had not electric motive power been discovered and developed. It was put to use on the old City and South London Railway, the first practical Tube railway in the world.

The development of the present Underground system from a number of separate railway undertakings into one unified whole may almost be described as a natural phenomenon, broadly applicable to other transport undertakings with long histories. London's system began with two pioneer underground lines, the Metropolitan (the world's first) and the Metropolitan District, which were progressively joined by eight others – all ten of them being private undertakings. A period of absorption and amalgamation culminated with all of them, plus other forms of public transportation, passing one way or another into public ownership and direction. Thus in 1948 we find the London Passenger Transport Board transferring its railway, bus, tram and trolley-bus systems, which it took over in 1933, to a newly-formed London Transport Executive; then in 1963 the London Transport Board assuming control; and in 1969 the Greater London Council taking over as policy maker and planning authority for all forms of transport, public as well as private (excepting that of British Railways) within its area.

Physically, London's Underground system penetrated less rapidly under central London than it spread as a surface network over the suburbs surrounding it. This too was a natural process as the 'catchment' area was far greater than the nucleus. The object in penetrating the then rural districts was to generate traffic, a highly successful exercise which resulted in large-scale housing development wherever 'Underground' lines were built. The costly business of distributing the resultant commuter traffic about inner London, largely by way of sub-surface and deep-level railway tunnels, continued as we have seen until 1906, after which only a relatively small amount of underground construction took place until the major work of building the Victoria line began in 1962.

Having somewhat briefly brought history up to date, one is left with the Victoria line, which is (or should be) the embodiment of all the past years' technological and operative skills. Outwardly there is not much basic difference between the Victoria Tube and its sixty-year-old predecessors. Unlike the San Francisco system, whose trains emerge from tunnels on to viaducts and are elegantly streamlined to speed over them, the Victoria line trains never emerge from their tunnels, except to ascend to the car depot at Tottenham. They are fundamentally tubular objects moving within a fixed tube, and there is not much scope for radical design change in such a concept.

But many new features and much complicated engineering have gone into or made possible this new railway, whose remarkable construction is largely buried in the earth, and whose ingenuous pieces of 'brain' apparatus are hidden away beneath seats or behind doors, so that nothing shows except smooth, unrevealing exteriors. The Victoria tunnels and their approach shafts, etc., lie beneath, or thread their way through a sub-strata of London which includes 'made' ground, pipes, conduits, sewers and railway tunnels that lie tangled and deep beneath London.

Under the Victoria main-line station the water-bearing in-filling of an old canal basin, over which the station was largely built, had to be chemically consolidated before mining could begin. At Oxford Circus a complex of five escalator shafts, two station tunnels and many connecting subways for the new Tube had to be fitted around the existing Bakerloo and Central lines' complex of tunnels. Here also the columns of an adjacent large store rest on the saddle of the new station roof, of exceptional strength, which safely supports one corner of the building.

To the north the Tubes roll one over the other so as to maintain a directional traffic flow where they run beside other tunnel tracks in Euston station; and north of King's Cross station they roll back again. At King's Cross itself the new tunnels become most involved with those of the Metropolitan Circle, the Piccadilly and the Northern lines, and with shafts to the sub-surface concourse, or ticket hall, that gives access to them all. One expects the unusual here and one finds it.

The notorious Fleet River, confined here in a sewer as big as a Tube tunnel, definitely intruded into the scheme of things; as did the Hotel Curve, a brick tunnel carrying trains from the Metropolitan

Above: Toronto: the waterfront; *below:* Toronto: train in open-cut
:tion of Bloor-Danforth line.

24. *Above*: Boston: city centre; *below*: Boston: streetcars of the Green
Line and train of the Orange Line at North Station.

25. *Above :* Chicago: the elevated Loop in 1919. The junction of Lake and
Wells Streets was once the busiest junction in the world; *below :* Chicago:
traffic congestion in State Street at Christmas 1964.

26. *Above*: Chicago: subway train in the State Street 'continuous platform' section of tunnel near Roosevelt station; *below*: Chicago: train of mode cars on the Congress Service (NW Route) in the median strip of t Express way.

widened lines tunnels up to the surface at King's Cross main-line station. The former had to be diverted to make more room for the concourse and the latter had to be carefully undermined where the new station tunnels run beneath. Thus there are trains running in tunnels at four different levels at King's Cross. The Victoria tunnels are, so to speak, tightly wedged between those of the Metropolitan and Piccadilly lines, and the Northern line tunnels run under them all, about ninety feet down.

The longer runs between stations on the Victoria line, together with the line's easy curvature and automatic driving, enable the new trains to run 20 per cent faster than those on other Tube lines. The 'get-away' in stations is faster too. A television screen on the platform alongside the operator's cab permits him to see all the platform, and as soon as all passengers are clear he closes the car doors, presses his twin 'starter' buttons, and from then on the train's progress is under automatic control. The trains work in accordance with a timetable code punched in plastic rolls in 'programme' machines, which automatically control signals and points, or switches: but the train's safety is looked after by the automatic control equipment. On the Victoria line the purpose and functions 'control' equipment and the driver 'command' equipment are in brief as follows:

Automatic Train Operation depends principally on electrical impulses or frequencies in the form of codes being received, interpreted and acted upon by apparatus installed in the train. All lines on the London Underground system, and on many other systems, are divided into sections for track circuiting; but on the Victoria line the state of the track circuits ahead, indicating whether the line is occupied, automatically determines which code of impulses shall be fed to the track in any section. A continuous series of impulses is transmitted through the running rails and picked up by inductive coils on the train. If no code is received the train cannot start; or should the train be in motion and the impulses cease for any reason, the train's brakes are applied to bring it to a stop.

The impulses forming these codes are produced, electronically or by pendulum action, at different frequencies of so many pulses a minute. Thus a code of 420 pulses a minute received by the train permits it to run with no restriction of speed: one of 270 pulses a minute allows the train to run under power at up to 25 mph, and one at 180 pulses a minute permits it to run at up to 25 mph providing power is not being fed to the motors. Train speeds permitted under the latter two codes are checked and regulated by apparatus on the train.

The foregoing is the safety signalling part of automatic train operation, and over-rides driver 'commands', which are given to the train through 'command spots'. These are sections of running rail about ten feet long, through which currents of a special frequency are passed. In practice, a train may be proceeding under power at unrestricted speed (under a '420 code') until it reaches a position where it can coast to the next station. At this calculated position there is a command spot with a 15,000 c/s current passing through it. Equipment on the train recognizes this and cuts off power to the motors. Nearing the station the train passes a series of 'spots' with speed-related frequencies, 100 c/s equalling 1 mph. The first spot with a frequency of, say, 3,000 c/s would bring the speed down to 30 mph, and then at following spots with decreasing frequencies the speed would gradually be brought down to that at which constant pressure braking brings the train to a halt at a predetermined place along the station platform. Apparatus on the train will, during the slowing down, have been comparing the train's actual speed with that of the frequencies received through the track and will have applied or eased the brakes as required.

All the running on this highly automated railway, which at the time of writing is operative between Walthamstow and Victoria, but is being extended beneath the Thames to Brixton, is under constant surveillance by two men in a control room near Euston station. Here a train regulator sees that trains run in accordance with the 'programmed' timetables, and a controller supervises the whole service including that in stations. He is aided by television which produces at his selection a picture of any station along the line, and besides instant communication with train operators he has of course an illuminated diagram of the route showing the position of every train along it.

It is reasonable to think that the Victoria lines' new features will provide the blueprint for any future Tube construction in London. So far as the latter is concerned, the only specific information at the time of writing is that London Transport is seeking Parliamentary powers to construct the first section of a new Tube, the Baker Street–Strand portion of the projected Fleet line that would cross London from north-west to south-east. If powers are granted and the necessary finance is forthcoming, London Transport hope that construction work will follow immediately after completion of the civil engineering work on the Victoria line Brixton extension.

The ultimate route of the Fleet line beyond the Strand would probably be via Ludgate Circus, Cannon Street, Fenchurch Street, and then under the Thames in a south-easterly direction. In addition to the Fleet line the Government has approved construction of a $3\frac{1}{2}$ mile westward extension of the Piccadilly line to Heathrow, to link London's main airport to the Underground railway system. There has also been mention of a scheme for British Railways to link the Northern City Tube line to the existing surface railways running out to places like Welwyn Garden City and Hitchin in the north, and to electrify the latter lines by an overhead current distribution system.

ADDITIONAL DETAIL

Authority: London Transport Executive.

Stations: There are 277 stations served by the system, including interchange stations which serve as many as three, and in one case, four underground lines. There are more than 200 escalators and nearly 100 lifts to serve the many deep-level stations, the deepest of which is Hampstead, 181 ft below ground. In central London they mostly take the form of two large diameter tunnels with side platforms in each.

Rolling stock: The newest cars operating on the Victoria line are 52 ft $9\frac{1}{2}$ in long (trailer cars are $4\frac{1}{2}$ in shorter), and 8 ft 8 in. wide; very slightly longer and wider than the older Tube stock. They have 'wrap-round' driving cab windows, and passenger compartment windows larger than normal to facilitate standing passengers' observance of station names. There is generally more standing room in the new cars, as the Victoria line caters for many short distance travellers on its 'in-town' sections. The cars on the shallower sub-surface lines, negotiating tunnels much larger than the deep-level Tube tunnels, are generally 9 ft 8 in wide and 12 ft 1 in in overall height, (about 2 ft 7 in higher than Tube cars).

Tunnels: Much of the system's sub-surface tunnelling dates from its early years and is brick-built, with rectangular tunnels 25 ft wide containing two tracks. The deep-level circular Tube tunnel, lined with cast-iron or precast concrete segments, is generally single-tracked and 11 ft 8¼ in in diameter. The Victoria circular tunnel is 12 ft 6 in in internal diameter, allowing for more clearance for speedier trains. The track gauge for all the system is standard 4 ft 8½ in.

Signalling: Basically automatic-block with two-aspect colour-light signals. The Victoria line's automatic operation largely dispenses with colour-light signals.

Power: Traction current is distributed by a four-rail system at 660v DC. Power to the system is generated in two London Transport power stations and fed to the system through 105 substations, all remotely-controlled.

Passengers: In 1968 the Underground carried 654 million passengers. This is fewer than on other large world systems but the average length of journey is much greater. On all the London Transport services (road and rail together) the figure for 1968 was 2,601,249,000.

CHRONOLOGY

January 10, 1863:	Opening of North Metropolitan Railway Company's Underground railway in London.
April 1870:	Opening of first seven-foot cable-car Subway (Tube) under the Thames.
October 1, 1868:	Opening of first Metropolitan District Underground railway in London.

December 6, 1869:	Original Thames Tunnel, completed 1843, first used by East London Railway trains (now part of Underground system).
December 18, 1890:	City and South London Tube opened with electrically driven locomotives.
July 30, 1900:	Opening of Central London Tube Railway.
March 10, 1906:	Opening of Baker Street and Waterloo Tube Railway (now Bakerloo Line).
December 15, 1906:	Opening of Great Northern, Piccadilly and Brompton Tube Railway (now Piccadilly Line).
March 7, 1969:	Opening of last stage of Victoria line between Warren Street and Victoria.

USSR

About one-sixth of the world's land area is Soviet territory, which contains a population of 234·4 millions. Within it there are nine cities with more than a million population, and several other cities likely to exceed one million population in a few years. There is a continuous growth in Soviet national economy, both in existing branches, in new types of production, and in the economic development of the remarkable riches in Siberia and elsewhere. Inevitably all this seems to point to the expansion of centres of population, and the creation of transport to serve them.

Up to the present, public transportation in the Soviet Union far exceeds private transportation, and even though the Russian attitude to private car ownership is changing and more cars are being built, public transportation in cities is everywhere favoured, and is not likely to be challenged by the private automobile in the foreseeable future. The expansion of existing underground-surface rapid transit urban railways (Metros), and the planning of new ones has been most marked in recent years – and it is reasonable to suppose that this process will continue. There is no knowing how many Metro systems the Soviet Union will possess in the next decade or so; but what is certain is that Russian engineers have acquired expertise in this field and have several times been called in to advise on building Metros, and to supply equipment for them, in major cities in other countries. In 1969 there were seven operative Metro systems in Russia, which are described briefly in the following pages.

MOSCOW

So far as the citizen's aspect of his particular Metro is concerned, it is as indispensable to him as the Underground is to the Londoner. In the early days of the Moscow Metro it was regarded as a privileged treat to ride the trains and pass through the Metro's splendidly ornate stations. As this Metro grew, from a single 7½-mile line in 1934 to four lines in 1950 and eight lines in 1969, with 87½ miles of route, so Moscow spread out its industry and built new estates of apartment houses on its perimeter. A vast commuter traffic thus generated was pleased to accept the technologically advanced new Metro, with faster trains and more business-like stations: but it still has warm regard for the old artistic Metro core, with its marble-halled stations and glistening chandeliers.

There is little doubt as to the essential role of the Metro in Moscow. The yearly total of passengers carried in 1960, 1,000 millions, is pretty considerable, but more so is the 1969 total of more than 1,590,000 millions. The average journey length on the Moscow system is much shorter than that on London's Underground, but the passenger total more than twice that of the London system's, does indicate a most intensive usage. (In 1964 the author was astonished at the size of the Metro crowds, even though he had just come from Tokyo.) The future shows little slackening in the growth rate of the Metro, to which thirty-three miles in new extensions and a new line are to be added in the next six years, and by 1980 perhaps a doubling of the 1968 total mileage.

ADDITIONAL DETAIL

The Metro operates under the authority of the Moscow City Soviet.

Stations: eighty-six stations, spaced on average just under a mile apart, are operative. Central stations are not only elaborate below-ground, but in some instances have palatial surface buildings. There are vast underground halls or circulating areas giving on to platforms, that at Komsomolskaya being about 107 yd long, 59 ft wide and 38 ft high. Marble surfaces predominate in these older stations. Escalators are installed at all stations where the depth warrants them.

Tunnels and track: Tunnels on the older stretches, built partly by the cut and cover method, are 25 ft wide. Later construction is mostly in deep-level tubes with an internal diameter of nearly 18 ft (5·46 m) which descend to as much as 131 ft (40 m) below ground at Dynamo station. Running track in welded rail is laid to 5·0 ft (1·524 m) gauge.

Rolling stock: The cars built at the Mytishchi Engineering Plant near Moscow are 62 ft 8 in overall and 8 ft 10 in wide (19·1 m by 2·7 m). They are made up into trains of seven cars (maximum). They have four automatically operated double doors each side, and longitudinal seating. All axles are motored. Cars built at the same works and of a similar type are in service on other Soviet Metros. Signalling and power: Basically the automatic-block, colour-light signal system operates, with speed control and train stops. Electronically controlled trains have been in operation on the Koltesvaya or Ring line since 1964 and are being progressively introduced on other lines. Traction current at 825v DC is distributed through a third rail by underside contact.

LENINGRAD

Leningrad, whose outer areas suffered heavily during its long siege by German forces during World War II, was the next Russian city to build a Metro. Since the October Revolution in 1917 Leningrad has grown almost four-fold in population, extensively rebuilding after the war and ringing itself with new residential districts and industrial projects. Its splendid palaces, museums, gardens and the like attract a small army of tourists every year – but intrinsically it is a busy, self-sustaining city, with a population of nearly 3½ millions which makes good use of its public transport, and particularly of its Metro.

The first section of its present Metro system, running northward from Avtovo to Vosstiana Square, was started in 1948 and opened in November 1955. An extension to this line was next added and it now connects with five main-line stations, the Moscow, the Vitebsk, the Baltic and Warsaw and the Finland stations. Between 1961 and 1963 a second north–south Metro line was opened, running from Victory Park in the south, through the city and beneath the River Lena to Petrogradskaya. In 1967 a third line, crossing the others and running east–west between Vosilioskrovskaya and Alexandra Nevsko Square, was opened. At the time of writing (1969) all three lines are being extended.

Here again in Leningrad the Metro is extensively used, though somewhat less than in Moscow. A total of 190 million passengers carried in 1963, when there was fifteen miles of route, had more than doubled in 1969 when 399 million passengers, each making a journey of just under a mile, were carried over its thirty-one miles of route.

ADDITIONAL DETAIL

Authority: Leningrad City Soviet.

Stations: Twenty-five stations were operative in 1969. Those of the earlier period include some that are richly decorated and some of more simple yet classical design. All are devoid of advertisement. Generally of the 'island' type with spacious halls or circulating areas giving on to platforms on either side.

Tunnels and Track: Leningrad is largely built on silty moist soil which does not favour shallow tunnelling. Consequently the Metro is built in circular tube about 17 ft in internal diameter, which descends to as much as 200 ft (60 m) below ground. Some of the construction has to be done under compressed air conditions. The track gauge is standard (Russian) 5 ft.

Rolling stock: Cars are built at the Mytishchi Plant, Moscow, and are similar to the Moscow cars. Trains are at present made up of four cars, but platform lengthening envisaged will allow for eight-car trains. Traction current is the same as in Moscow – and like the Moscow system, the Leningrad Metro is progressively introducing automatic train operation.

KIEV

The Ukranian capital, with a population of about a million and a half was the third Russian city to possess a Metro. Its long history reached a climax in World War II when its centre was battered by enemy action, and not a building higher than one storey was left standing on the Dneiper's left bank. When the war ended, after a long occupation by German forces, Kiev's cultural and social life began again with the restoration of the fine buildings and wide streets of this city. Now it has overstepped the old boundaries and big residential districts have been built on the outskirts. Although these possess their own schools, hospitals and shopping centres, they need mass transportation to link them one with the other, and to the centre. It is to meet this need that the Kiev Metro, consisting presently of one line, is being developed into a three-line network reaching out from the centre to the suburbs.

The first three and a half miles with five stations, from Terminus Station north-east to Dneiper, was opened to passengers in November, 1960. A major engineering feat was the crossing of the River Dneiper by a bridge that carries motor traffic and Metro trains on different levels, and whose Dneiper station is built on the bridge approach. By 1964 the crossing had been accomplished and the line opened up to serve vast housing estates at Nikolsk and other north-east suburbs.

ADDITIONAL DETAIL

Authority: Kiev City Soviet.

Length of operative Metro route: (1969): 9·7 miles (15·8 km), of which 5¼ miles (8·5 km) is in tunnel and 4·4 miles (7·3 km) is above ground. About 2½ miles of line in tunnel is being built to extend the first line, which ultimately will extend from north to south of Kiev for a distance of 15 miles (24 km). The first section of Kiev's second Metro line was begun in 1969. It will be nearly six miles long, all in tunnel, and will make a junction at Kretschatik, Kiev's main thoroughfare, with the first line.

Stations: The deepest is Arsenal, 328 ft (100 m) below ground, which is reached by two banks of high-speed escalators. Terminus Metro station forms part of the principal main-line station. Generally the inner stations lie at very deep level, but this does not detract from the high quality of architectural decoration, which employs Ukranian granite, coloured marbles, majolica and anodized aluminium.

Tunnels and track: Under the city the deep-level tunnels are circular, with an internal diameter of 16 ft 9 in (5·1 m), containing a single track of normal Russian 5 ft gauge. Construction methods include excavation by rotary cutting machinery.

Power and signalling : Traction current at 825v DC is distributed by third rail. The automatic-block signalling system with colour-light signals and train stops is basic, but at terminal stations there is also automatic reversing equipment.

Rolling stock : Cars similar to those in use on the Leningrad and Moscow systems (from the Moscow Mytishchi Plant) are made up at present into three-car trains, but five-car trains are contemplated. They run at two-minute intervals during peak periods.

The total of passengers using the Kiev Metro during 1968 was 103 millions, illustrating an intensity of usage equal, when related to the operative length of line, to that of the Moscow Metro. There are also efficient services of buses, trolley-buses and trams.

TBILISI

Next in order of time comes the Metro in Tbilisi, which this capital city of the Georgian Republic first acquired in 1965. Tbilisi (the name derives from hot curative springs discovered here in the distant past) has developed in recent years into the administrative, cultural and scientific centre of the country, and now has a population approaching one million. It stretches along the banks of the River Kura, and large residential and industrial areas are located at either end. It was to link these areas with the centre, and provide existing commuters with a swift, mass transportation facility such as an underground railway affords, that the Metro was built.

The first section of line, 3·75 miles (6·0 km) long, connecting the Didube and the Twenty-six Commissars regions (Komissarov) was opened for traffic in 1965. This line has since been extended about two and a half miles and a future spur line is envisaged.

The Tbilisi Metro, which operates under the authority of the City Soviet, was designed by engineers of the Caucasus Design and Survey Institute. Excavation of the shallow running tunnels and of the station sites was undertaken from the surface. At the time of writing (1969) the line runs for 6·4 miles (10·3 km) between the Didube and 300 Avagvintsev stations. There are nine stations, seven of which are underground. The system's main features, which include its rolling stock, signalling, track gauge and voltage and distribution of traction current, all conform to standard Russian practice (that is, their detail is the same as that listed under the Moscow Metro).

The city's public transport system comprises the Metro, trams, buses and trolley-buses. Metro passengers in 1969 totalled 68·3 millions.

BAKU

This capital city and seat of government of the Azerbaidjan Republic, lying on the west coast of the Caspian Sea, had a population in 1968 of 1,280,000. It has come into prominence in the last few decades through the exploitation of oil fields in the area, and by industries deriving from that exploitation. Oil and gas were discovered there in ancient times. Industry, however, has not detracted from Baku's popularity as a seaside resort, and the description 'a white city arranged in terraces on the gently sloping hills around Baku Bay' is apt. Lying offshore there is another city, Neftyaniye Kamni, built on piles in the bay. It is made up of the many drilling rigs, shops, housing apartments, schools, etc., that are linked together by 170 km of trestlework and also linked to the shore by a highway. Its business is concerned with the getting of the oil.

The Baku Metro was Russia's sixth. Its first and presently-operating sections of line totalling 6·2 miles (10 km) in length were opened for traffic in late 1967. They comprise a transverse line and a spur, with six stations, four of which are at deep level and are reached from the surface by escalators. Variations in the ground through which the tunnels were driven ranged from rock to quicksand, necessitating the use of caisson construction methods in places. As with Tbilisi (400 miles to the west) the Baku Metro employs the standard types of Russian rolling stock, signalling and traction current voltage and distribution method. It operates under the authority of the Baku City Soviet, who plan to extend the system east and westward from 28 April station, bringing the total number of stations to eighteen.

KHARKOV

The second city of the Ukraine, with its satellite towns, contained in 1968 a population of between 1¼ and 1½ millions, and extended over an area whose radius from the city's centre was twenty miles. The extensive services of buses, trams and trolley-buses serving this area are proving inadequate to meet present and future demands and the first part of an ambitious system of underground railways, planned a few years ago, is now under construction, and scheduled to carry its first passengers in 1974.

On this system, underground trains will interwork with suburban trains, under the city and beyond it to the outskirts, similarly to the system operating in Tokyo. Tunnels are being built to an internal diameter of 20 ft to accommodate both types of train, which will derive their traction current from an overhead wire distribution system at 3.000v DC. Suburban trains will ultimately run from one outer area to another, traversing the city underground.

Rolling stock, of both urban and suburban commuter type, being built at the Riga coach building works, will be capable of speeds up to 56 mph (90 km/h) underground, and 75 mph (120 km/h) on outer stretches. The first east–west link, which will run for about 6½ miles underground, will be served by eight-car trains made up of four-car units. Station platforms to take such trains will be 328 ft (100 m) long. Construction of a second stage of this urban-suburban system is expected to provide another 14 stations and 11¾ miles of route. A third stage construction project takes the form of a spur line to serve future housing development east of the city.

TASHKENT

The capital of the Uzbek SSR, in developing plans for what will be the seventh Soviet Metro, will conceivably not be the last in the Soviet Union to do so. It has been announced that Minsk, some 2,000 miles to the north-west of Tashkent, is considering building a twenty-mile Metro.

Tashkent has over the last decade or so increased remarkably in size and importance, and in 1968 had a population of 1,324,000. The projected Tashkent Metro will have as its first priority a line to serve regions of mass housing and industry, in being and being built. (In part this is rehabilitation following the earth-quake of 1966). There are to be twelve stations on this first 16·2 km line, which will traverse the city from north-east to south-west. According to a master plan, future second and third Metro lines would add 50 km to the system. The design of the first line has been completed.

MONTREAL

The first underground railway in the world, in London, came into being primarily as a result of one man, a City solicitor (or lawyer) named Charles Pearson, who conceived the basic idea. Something similar is recorded of Canada's second underground railway project, the Montreal Metro; although here it was a case of advocating an established transportation system for a modern city. Several earlier suggestions for a Metro for Montreal, made over a very long period, had no positive outcome, although the city by virtue of its growth and importance (it is Canada's largest city with a population of more than two and a quarter millions) has surely qualified for one. Even so, it was not until 1961 that the City of Montreal, largely as a result of the advocacy of its present Mayor, Mr Jean Drapau, authorized the building of Lines 1 and 2 of its present Metro. Construction work began on May 23, 1962, and on October 14, 1966, just over ten miles of these routes were opened for passenger traffic.

In the meantime, authorization had been obtained for Line 2 to be extended at both ends and for a line designated as No 4 to be constructed. A proposed No 3 Metro line has still (1969) not received the 'go-ahead', the break in number sequence being explained by the following: The No 3 line proposal was to have utilized the Canadian National Railway's electrified route between Central Station in downtown Montreal and Cartierville, about nine miles to the north-west (the line that tunnels through the 770 ft Mount Royal, dominating the city). In 1962, however, permission was received for the next World Exposition to be held in Montreal in 1967, and the sites subsequently chosen for this were on islands in the St Lawrence River.

This important event necessitated an immediate change in Metro planning. A fourth line was authorized in 1963 to run from the junction of Lines 1 and 2 at Berri de Montigny, south-eastward under the St Lawrence River to St Helen's Island and then under the main channel or Seaway to Longueuil on the far shore. Thus (if it could be built in time) it would provide adequate mass transportation between the city and the Expo 67 site, and subsequently be extended to developing areas beyond Longueuil.

As history will no doubt record, this complicated underground and under-water tunnelling project was in fact accomplished in less than three years, and opened for traffic about a month before the official opening of 'Expo 67', which took place in the Spring of 1967.

The topography of Montreal and the plan of its Metro lines are remarkably similar to those of Toronto and its Subway, described also in this book. Both cities have a chequer-board pattern of streets and their busiest districts are those lying near their respective water fronts. Montreal's No 1 line was not built directly beneath its main east–west thoroughfare, St Catherine Street, which serves the Universities and shopping districts. To do this would have meant a closure or partial closure of this busy thoroughfare and dislocation of its considerable surface traffic during the construction period. Therefore the line was located beneath minor paralleling streets; and the same principle was applied in locating the north–south Line 2, which closely parallels St Denis Street. It will be seen that a similar precedure was adopted for the downtown lines in Toronto.

The Montreal system does, however, differ in main features from any other system, with the exception of the Paris Metro. It was built to serve the second largest French-speaking city in the world, and was in fact modelled in many respects on the Paris Metro, whose officials (of the Régie Autonome des Transports Parisiens) were consulted on its design and operation. Thus the pneumatic-tyred type rolling stock operating on the Paris Metro was adopted for the Montreal Metro. A description of this type of stock is given in the Paris Metro reference, but here it may be said that the adhesion-to-track qualities of the rubber tyres do enable the Montreal trains to negotiate without difficulty the steep switchback profile of tunnel lines on each side of the stations, a construction feature which assists trains to accelerate rapidly on leaving stations and to decelerate on approaching them.

27. *Above*: Chicago: articulated train on the portion of the Skokie Swift route equipped for overhead current collection; *left*: Cleveland: city centre.

28. *Above :* Cleveland: one of twenty new cars for the Airport extensio
centre : New York: Third Avenue elevated railroad before electrificatio
about 1880; *below :* New York: the latest air-conditioned R42 type c
on New York Subway (1969).

29. *Above :* New York: night scene in mid-Manhattan; *below :* New York: stainless steel rolling stock on the former BMT division of the NYCTA.

30. *Above:* New York: latest cars for the NYCT with moulded fibre
glass ends; *below:* New York: stainless steel rolling stock on the former
BMT division of the NYCTA.

As with the rolling stock, so with the stations on the Montreal system, which can be likened to modern conceptions of those on the Paris Metro. Unlike Paris, there are no conventional sidewalk entrances to stations. Access is generally through covered-in structures or buildings; and in this connection one sees evidence of the growing trend in major world cities to construct underground shopping centres, or commercial complexes, and to connect them to, or build them as integral parts of transportation centres. The Montreal complex, linked to the Metro, at present covers fifty acres and is planned for expansion to double this size. One feels that the somewhat futuristic conception of underground cities may not, after all, be so far away.

ADDITIONAL DETAIL

The Montreal Metro is the property of the City of Montreal, which financed its construction by way of loans, appropriating credits totalling $213,700,000 for construction, trackwork, rolling stock and accessories. The city's Public Works Department had charge of preparation of plans, calling of tenders and supervision of work.

Route length (1969): 16·13 miles (26·0 km) all underground.

Passengers carried in year ending March 1969 by the Montreal Transportation Commission (which operates the Metro) services, both bus and rapid transit: 287,000,000, of which 125,000,000 were carried on the Metro.

Maximum train frequency: Twenty-four trains per hour on Line 2.

Tunnel and Track detail: Single tunnel, mostly in rock, carries two tracks. It is 23 ft 4 in wide with vertical walls and arched roof and lies at depths varying from 20 to 180 ft. below ground level. About 30 per cent was built by the cut and cover method. Standard 4 ft 8½ in track is required by the pnuematic-tyred traction system.

Maximum gradient: 6·3 per cent.

Rolling stock: Cars manufactured by Canadian Vickers Ltd. Montreal, are 56 ft 5 in (17·2 m) long. (motored cars) and 53 ft 10 in (16.5 m) long, (trailer-cars). Both motor and trailer cars are 8 ft

3 in (2·5 m) wide. Nine cars semi-permanently coupled in three-car sets form a train. Two men per train act as driver and conductor, or door operator; each man being stationed in a control department at either end of the train, changing his function at the end of each run.

Stations: All have side platforms 500 ft long. Each differs from the other in architectural detail, the effect aimed at in all being to create an aspect of spaciousness and daylight, the latter by employing indirect artificial lighting.

Traction current and signalling: Traction current at 750v DC is fed to the trains through lateral guide bars each side of track. Signalling employs the automatic block system with track-side signals.

TORONTO

Toronto, capital of the province of Ontario, has the distinction of possessing the first urban Underground or Subway in Canada. As far back as 1910 when there were less than 380,000 people living in Toronto, the city was reported as having a traffic problem, and to have put forward a suggestion for building subways, presumably to relieve traffic congestion. In subsequent years there were other propositions relating to the need for a form of rapid transit, including one for putting the city's street cars into tunnel beneath the downtown area; but in the event, nearly forty years elapsed after the original proposition before a definite decision was taken to build a conventional subway line. There may have been cogent reasons for this long deferment, not the least being (one feels) the increase in ownership of private transport and the general effect of 'motorization' on development and on the public's travelling habits.

Toronto's subway system today is made up of two routes along its two main traffic arteries, which run north–south and east–west, as do nearly all the streets of the city. Until the advent of the first subway, Toronto had one of the largest street-car systems in the world. Yonge Street, the city's busiest north–south thoroughfare, was also a very busy street-car route where at peak periods street cars followed one another at less than minute intervals. It was obvious that this route should be one of the first chosen for a subway, and one was built, not directly beneath the street but mostly on a near parallel alignment (so as not to cause undue interference with surface traffic during construction) and named the Yonge Street line. It runs from Front Street in the downtown area (its nearest point to Lake Ontario) northward for 4·6 miles to Eglinton. For about three-quarters of its length the line is in tunnel built by the cut and cover method, and in open cutting for the remainder, mostly at its northern end. The subway carried its first passengers on March 30, 1954.

Toronto's second main subway was built on an east–west axis, following the line of Bloor Street and its easterly continuation, Danforth Avenue. These thoroughfares were, like Yonge Street at one time, congested with a mixture of automobiles and closely-spaced street cars, and this second route was ultimately decided on for a subway. The Bloor-Danforth line was opened for passenger traffic between Keele station in the west and Woodbine station in the east, on February 26, 1966. Of its total length of eight miles, nearly all is in cut and cover tunnel.

An extension of the Yonge Street line, paralleling it at its busy downtown end and thus relieving it of heavy traffic at this end, was opened in February 1963. It runs under University Avenue, extending the Yonge Street line by 2·38 miles, all underground but partly in shield-driven circular tunnel. At its northern end the University extension forms a junction with the east–west Bloor–Danforth line.

The most recent extensions to the Toronto system are those at either end of the latter line. They were opened on May 11, 1968, and added a further six miles to the Bloor–Danforth route, which now terminates at Warden in the east and Islington in the west. These extensions were variously in tunnel, on embankment, and over long concrete bridges in the west where the line passes over a river and a creek. Costwise the construction and equipping of these extensions, totalling $77,723,000, is being met by the Municipality of Metropolitan Toronto as to approximately $60 millions and the Toronto Transit Commission (which operates the Subway) approximately $17½ millions. By this arrangement or agreement the Commission was enabled to proceed with construction without substantially increasing fares.

Under construction since September 1968 is a further extension of the Yonge Street line which will carry it four miles northward into the Borough of North York and add three more stations. It is scheduled for completion in 1972 and is expected to cost $79 millions.

There are two items of particular note about the Toronto system. The first may equally apply to other newly-built rapid transit lines, but has been emphasized in respect to Toronto, where the Subway has had the effect of considerably enhancing land and building values. The second is Toronto's introduction of subway cars equalling in length those on surface railroads. By increasing the car length to 74 ft 5⅝ in (the longest of any at one time*) the amount of under-car equipment per passenger carried was reduced, as was initial cost and car weight, the latter being further reduced by the use of aluminium extrusions for structural elements.

ADDITIONAL DETAIL

The Toronto Transit Commission, which operates the city's street cars, buses and trolley-buses, as well as the Subway, serves the Metropolitan Toronto area, which has a population of around two millions. In 1968 it carried approximately 324,000,000 passengers, of which total 147,000,000 were carried on the Subway. Over recent years the Subway's patronage has progressively increased, as illustrated by the following (rounded off) totals: 1961, 66 millions: 1964, 76 millions: 1966, 117 millions.

Route length of the Toronto subway system: (1969) 20·73 miles (33·2 km) of which 17·44 miles (27·0 km) is in tunnel.
Number of Stations 1969: Forty-five. There are both side and centre platform type stations (platform length 500 ft). The deeper stations have escalators. No two adjacent stations have the same colour scheme, a feature which assists passengers to recognize their stop.
Track: The subway track gauge of 4 ft 10⅞ in (1·5 m) is unique in the rapid transit world. It was adopted to conform with the street-car track gauge and enabled subway cars to reach a street-car depot for certain maintenance.
Tunnels: Mainly of rectangular section, they were built at a minimum of 8 ft (2·5 m) below ground to afford frost protection They are double-tracked and 32 ft 6 in (9·9 m) wide.
Rolling stock: This comprises older stock built by the Gloucester

* The overall car length for the Washington Metro prototype car and the Paris Regional Express cars are 75 ft and 72·9 m per 3-car set (approximately 79 ft 8 in per car) respectively.

Railway Carriage and Wagon Company, England, and newer stock built by the Montreal Locomotive Works and Hawker Siddeley Canada Ltd. The newer cars, 74 ft 5⅜ in long, are approximately 17 ft longer than the older cars. The passenger capacity of a six-car train of this newer stock is equivalent to that of an eight-car train of older stock.

Signalling: A three-colour signalling system, with train stop mechanism, works on the automatic block principle.

Traction current: The system operates on 600v DC traction current distributed by a third rail.

Service Frequency: The maximum number of trains per hour in each direction is 26 to 27.

Car maintenance: This is carried out in car depots and workshops at Davisville and Greenwood.

BOSTON

Boston, capital of Massachusetts, is situated at the junction of the Charles and Mystic rivers about 215 miles north-east of New York. It was founded in 1630 by the English Puritans and named after the town in Lincolnshire, England, from whence many of its first inhabitants came. During all its history Boston has been renowned as a centre of learning (Harvard College, the first in the USA, dates from 1640), but the city also has a long maritime history, besides being the birthplace of American industry.

The narrow streets in Central Boston, retained from its distant past, have always posed traffic problems, and these were accentuated as metropolitan Boston grew in area by the inclusion of neighboring communities within its boundaries. As early as 1897 it was found necessary to relieve the central streets of some of their surface traffic by building a tunnel and diverting street cars into it, and this process continued until there were no more surface street car routes in Central Boston. These routes and others are now operated by modern types of street car (locally called trolley-cars), and since they run partly in tunnel and partly in their own rights of way on the surface, the distinction between them and conventional underground trains has lessened.

There are, however, three clearly defined rapid transit railway systems operating in Boston today, and they are planned for considerable extension in the future. The first to operate was the 'Elevated' railway, now known as the 'Orange' line, from Everett in the north to Forest Hills in the south. This railway, opened in 1901, runs partly in a tunnel 1·8 miles long, with four underground stations beneath the city, and so by its age and route qualifies as the oldest 'underground' in the west. The second is the Harvard Square-Ashmont line from west of the city to south-east, now known as the 'Red' line, which opened in sections between 1912 and 1928 and also runs in tunnel for part of its length. The third is the East Boston line, now known as the 'Blue' line, which runs in tunnel from Bowdoin in the city centre, eastward under Boston Harbour in tunnel which was first used by street cars, and thence on the surface to Wonderland in the north-east. In all, these lines total twenty-three miles of route, about half of which is underground.

The population of Boston in 1967 was about 700,000, not large by American city standards, and not expected to show any dramatic increase in future. Its public transport services, however, are extensive and very mixed, comprising elevated and underground railways, surface and underground street cars (which operate through five miles of special street car tunnel), buses and trolley buses.

In 1964 the concept of public transportation for Boston and its surrounding country underwent a radical change when the Massachusetts Bay Transportation Authority came into being. This representative Authority is charged with the co-ordination, expansion, improvement and operation (as well as financing) of public transportation over all the area within an approximate twenty-mile radius of Central Boston; an area which embraces seventy-nine cities and towns and contains a population of nearly three millions. The exacting nature of the Authority's task lies in providing a system of public transport attractive enough to compete with the motor car, and in endeavouring to satisfy the transportation needs of numerous self-governing communities in such a way that the needs of one are not met at the expense of another. The comprehensive programme of proposed improvements and extensions made public two or three years ago would, if completed in their entirety, have cost more than $300 millions: but since the announcement of the original programme it is reported that a new master plan is under consideration for proposed expansion that would considerably increase the rapid transit extensions, and of course relatively increase their cost. In view of the long-term nature of the various projects and possible changing requirements as time goes on, it is reasonable to expect that plans would be subject to review from time to time. Some of the original proposals are given below in brief outline:

A Extension of the Cambridge-Dorchester line (the Red line) 2·5 miles north-west from Harvard.

B Re-routing of the Everett (Orange) line, via a tunnel under the Charles River, and the ultimate removal of all the elevated sections, plus a south-westward extension of the line from Forest Hills to W. Roxbury.

C Construction of a new rapid transit route from South Station in Boston south-eastward for fifteen miles (the proposed new South Shore line).

D Upgrading of the high-speed street-car line from Ashmont southward to Mattapan, so that rapid transit trains could continue through Ashmont to Mattapan.

At the time of writing work is proceeding on the construction of the double-track Charles River tunnel for the Orange line's new route, and on a bridge over the Neponset River for the South Shore line, for which seventy-six new cars have been ordered. Other works too numerous to mention in detail are under way, in connection with the Everett line's new construction, the South Shore line, and general modernization.

ADDITIONAL DETAIL

All the lines, of 4 ft 8½ in (1·435 m) track gauge, are double or multiple tracked.

Stations: There are forty-one stations, several of those in the tunnel sections being equipped with escalators.

Tunnels: These, with the exception of the under-water tunnels, were mainly built by the cut and cover method, being rectangular in section and of double or multiple track width.

Rolling stock: The latest cars on the Cambridge (Red) line are 69 ft 9¾ in long and 10 ft 3½ in wide, considered amongst the largest until the Toronto Subway introduced even longer cars. the cars on all lines are motored and have a capacity of from 278 passengers, including 72 seated, on the largest to 178, including 44 seated, on the smallest.

Traction current: 600v DC. There is a third rail system on all lines

with the exception of the outer four miles of the East Boston (Blue) line to Wonderland, which has overhead wire collection.
Signalling: The rapid transit lines employ the automatic block system with colour-light signals and train stops.
Workshops: The main overhaul shops are at Everett, with car depots at four other locations.

In 1968 the Massachusetts Bay Transportation Authority carried 193 million passengers on all its services (rapid transit, buses, street-cars).

CHICAGO

Chicago's location on maps of the USA shows clearly why it has been described as the 'city at the cross-roads of a nation'. The early settlers in the 1830s, bound for the Pacific coast 1650 miles away, or even farther to the North-west Territories, chose this spot to stay over whilst replenishing stores. Chicago became settled in this way, as a trading post and stop-over for the biggest migratory movement in the country's history, and thus began its own history that throughout has been linked with transportation, through the whole range of wagon trains, stage coaches, railroads, shipping, and lastly auto-highways and airways.

With some twenty major and sixteen minor railroads centring on the city, plus a network of motor roads (including the sixteen-lane Dan Ryan Highway), dozens of air routes and not a few shipping routes, Chicago appears to the visitor to be the traffic hub and economic magnet for all North America. To attempt a description of the city's growth without superlatives would be difficult, but little less would suffice to convey a true impression of Chicago's remarkable expansion, in about a century and a half, from a settlement to a vast metropolitan area with more than seven and a half million inhabitants. This has been due not only to its strategical position, but also to its capabilities for producing food in immense quantities, and most other manufactures from a great surrounding area that yields prolifically from its fertility and mineral richness.

About one-third of Chicago is taken up by streets, by-ways, highways, railways and waterways. It was laid out on a grid pattern mostly before the advent of the motor vehicle, and today this traffic not only streams along the expressways but filters into the small streets, and particularly into the busy 'Loop' or centre area of towering office and apartment blocks. Inevitably traffic clogs the busiest streets, but steps are being taken (they are referred to later) to mitigate congestion, a condition that should noticeably improve after two new rapid transit lines open for rail passengers in 1969, and the prospect of weaning commuters away from their autos becomes that much brighter.

The beginning of Chicago's rail rapid transit was in 1892, when the Chicago and South Side Elevated Railroad operated its first trains over elevated urban tracks. The renowned Loop was built five years later. This is a rectangle of elevated railroad like a traffic roundabout, enclosing a square mile of the busiest portion of downtown Chicago. At one time its tracks were the most densely operated in the world – and may still qualify for that title while the structure lasts. It provides the rail passenger with an excellent elevated view, clear of jostling traffic below, of the interesting Chicago River activity and general city life, and many a commuter will miss this ride when the Loop is pulled down. Its removal has been firmly proposed under a comprehensive traffic re-organization scheme, which replaces the Loop by a Subway (underground railway) and also envisages rapid transit railroads, ordinary railroads and highways all in single traffic channels.

Elevated railroads were built radially from Chicago's centre, which later became the Loop, in the last years of the nineteenth and the first years of the twentieth centuries. The first Subway, the 4·9 mile State Street line, linking the North Side Elevated system near Armitage Avenue to the South Side Elevated system near 16th Street, was not opened until 1943. So for fifty years the city relied principally on its elevated system, which was electrified almost from the beginning, first through locomotives and then through multiple-unit stock, and provided frequent and reliable services which often were carrying record numbers when other surface transport was bogged down in snow blizzards.

The second Subway, the Congress line, was opened in 1951 for 3·9 miles, and extended in 1958 (in tunnel for three quarters of a mile and then as a surface line) to form a U-shaped transit line linking western Chicago with the north-west via the Loop district. The 1958 extension prolonged the southern end of the line in a westerly direction along the Congress Expressway, and was the first instance of rail rapid transit routed along a multi-lane highway in a grade-separated right of way. The trains run in open cutting between the traffic lanes. At its north-western end, the Congress line (now known together with the Cicero service as the North West Route) ascends from tunnel to join the Elevated tracks and its trains continue along these to Logan Square. Thus Chicago has relatively little rapid transit line actually in tunnel at the present time.

An unusual feature on both Subway lines is the 'continuous platform' in the tunnels on their downtown sections. On the State Street line the platform, about three-quarters of a mile long, is connected to eight sub-surface ticket halls but has only three train stops along its length. The Congress line's platform is similar, but is about a quarter of a mile shorter and has six sub-surface ticket halls.

In 1964 the Chicago Transit Authority made an interesting experiment when it purchased a five-mile stretch of surface line in the Skokie district (to the north of Chicago) from the Chicago, North Shore and Milwaukee Railway. It was rehabilitated between Howard and Dempster stations, equipped with standard CTA cars converted for overhead current collection, and generally brought up to rapid transit status to act as a feeder line to the main North South rapid transit route. Its trains run non-stop between Dempster and Howard in six and a half minutes, and the line, renamed the Skokie Swift, has since been reported as being completely successful and attractive to commuters, many of them auto owners who leave their autos in the large park, or lot, at Dempster station.

In 1966 or thereabouts the comprehensive plan for Chicago referred to earlier was made public. Its proposals range over a very wide field and are partly concerned with facilitating movement generally within the city. The placing of two or even three different types of transportation in one traffic corridor is part of the scheme, and a start in this direction had already been made when the Congress rapid transit line was routed along the middle of an Expressway. The two new rapid transit extensions, which will be operational by the autumn of 1969, follow this principle in so far as they share a highway with other traffic.

The Dan Ryan Highway project is a 9·5 mile rapid transit extension to the CTA elevated system in the south. It starts near 17th

Street and continues southward along a median strip of the Dan Ryan Highway. The other is the Kennedy extension in the northwest, where a one-mile tunnel from Logan Square conveys rapid transit subway trains to the median strip of the Kennedy Expressway, along which they continue for four miles to Milwaukee Avenue. The total reported estimated cost of the two extensions and its new cars is $107 millions, part of which is a Federal government grant. The extension lines will eventually be connected to the new Subway system replacing the elevated Loop Tracks, but the latter project is not expected to be operational until 1974.*

ADDITIONAL DETAIL

Authority: Chicago Transit Authority, which also operates the buses and trolley-buses.

Length of route: The system, which includes the two Subway lines and six elevated lines, totalled in 1969 71·13 miles (114 km) of which 9·1 miles is in tunnel: but to this total will shortly be added the tunnel and surface route mileage of the two new extensions.

Tunnel and track detail: The tunnels include sections of tubular and cut and cover construction, the former lying generally some 40 to 50 ft below surface level. The track gauge is 4 ft 8½ in (1·453 m).

Stations: Generally these are approached by stairs from the street through sub-surface ticket halls, but escalators are provided for ascending passengers.

Signalling: The basic system is automatic block with three-colour light signals and train stops, but the cab-signal system with speed control apparatus already in use on one surface line is being installed on the new extensions.

Traction Current: At 600v DC over almost all system through a third rail system. On the outermost section of the Evanston branch and on part of the Skokie route, by overhead wire collection.

* The Dan Ryan and Kennedy Expressways Rapid Transit extensions were opened in September 1969 and February 1970 respectively.

Rolling stock : To the total of about 1,140 cars owned by the CTA in 1969 will be added those bought for the Kennedy and Ryan new routes. They will be air-conditioned, the same as are the 180 cars delivered in recent years for the existing system.

In 1969 the whole CTA system carried about 420 million passengers, of which nearly 104 millions were rapid transit rail passengers. Although the Chicago system attracts less passengers mile for mile than the New York system, the trend is towards more rapid transit coverage for Chicago and increased patronage. The fare structure is based on a minumum fare with surcharges for travel to and from certain outlying stations.

CLEVELAND

Cleveland, Ohio, with a population of approximately one million, lies on the south shore of Lake Erie, 350 miles east of Chicago. It is a comparatively young city, founded in 1836, with a history over the last 100 years of rapid industrial expansion, and population growth that led to its present position as the seventh largest among American cities. Cleveland's geographical position on trunk railroad routes between the east and west coasts was a factor in its growth, but primarily this was due to the western world's need for iron and steel products, which Cleveland has been able to meet on a large scale. It is in a favourable position to do so since it has the necessary smelting agents, coal or oil, within easy land distance, and iron ore from the regions of Lake Superior is transported relatively cheaply and directly to Cleveland by bulk carriers across the Great Lakes.

Cleveland's urban public transport needs were met originally by several private street car lines (to which buses were subsequently added). They were acquired as a system by the City in 1942. For some years prior to this date proposals had been made for an urban rail rapid transit line, capable of accommodating the flow of commuter traffic entering the city principally from its eastern suburbs; and the rights of way for such a line were in fact basically reserved in the 1920s. The depression years and World War II postponed any favourable decision that might have been reached, and it was not until 1952 that work began on a thirteen-mile rapid transit line, made possible by a loan to the City, advanced by a financial agency to the Federal Government, of sufficient capital to build the line, improve surface facilities and buy new rapid transit cars.

The line, between Windermere in the east, via Union Railroad Terminal at the city centre to West 117th Street, was opened in 1955. It was extended two stations westward to West Park in 1958.

31. *Above*: New York: interior of new air-conditioned Port Authority
Trans-Hudson (PATH) cars; *below*: Philadelphia: stainless steel cars on
the Market-Frankford Subway-elevated line.

32. *Above:* San Francisco: down
town district; *right:* San Francisco:
tunnel excavation for Line 3 under
Oakland (on mainland).

33. *Above:* San Francisco: model of new station on Bay Area Rapid Transit Railway; *centre:* San Francisco: cut-away of ground below Market Street, showing Bay Tube at lowest level and new street-car tunnel above; *left:* San Francisco: giant catamaran barge lowers one of fifty-seven tube sections into place in the four-mile Trans-Bay Tube between Oakland and San Francisco.

34. *Above :* San Francisco: mock-up of the Bay Area's modern rapid tra⟨n⟩
sit car; *below :* Washington: architectural model of typical side-platfor⟨m⟩
Station of the future Metro.

Only a short section of the whole line, that part which runs beneath the Union Terminal, is in tunnel. The most recent addition, completed in 1968, is the important surface extension of 4·08 miles (6·8 km) westward to the Cleveland Hopkins International Airport.

There is also another rapid transit system in Cleveland, the Shaker Heights line. This line was built in the 1920s to develop the latter community and was taken over by the City of Shaker Heights in 1944. It is operated with high-speed street-car type vehicles. The line runs eastward from Union Terminal, sharing the same tracks with the Cleveland Transit line (and the same overhead power supply) for 2·55 miles to East 55th Street. From thence it branches south-east and divides, one line continuing to Shaker Heights and the other to Warrensville Heights. It has a total route length of 13 miles (21 km).

The Cleveland Transit airport extension constituted a milestone in rapid transit history as the first direct rapid transit line into an airport in the USA. The airport terminal station is beneath a car park at the port's eastern entrance and is only a minute's walk from the airport check-in counter. Two new rapid transit stations are along the surface extension, which was estimated to have cost $18·6 millions, plus approximately $3½ millions for twenty new cars for the extension. The stainless steel cars built by the Pullman-Standard division of Pullman Inc. are 70 ft long, driven by 100 hp motors giving a maximum speed of 60 mph and are air-conditioned. Electronic equipment in the driver's cab picks up commands transmitted through the track, and permitted speeds are indicated on a cab display panel. Incorporated in the system are route-setting devices operating under remote control. The system is fail-safe, the train automatically stopping if continuous communication is not received.

The Cleveland Transit line as a whole does not tap densely populated areas and depends largely on feeder bus route passengers, and patrons with private cars. It has therefore concentrated on providing extensive car parks at its stations. (The two intermediae stations on the airport extension each have parks for 1,300 cars). Possible extensions to the line were outlined in a recent report submitted by the Seven County Transportation–Land Use Study group. It recommends considerable expansion, adding seventy miles of rapid transit route, including sixteen miles of subway, to the Cleveland Transit and Shaker systems. Extensions would prolong the airport extension westward, and the Shaker lines east and south-eastward. The expansion scheme, estimated to cost in all $570 millions, would also include new radial lines from down town Cleveland, and construction of an underground distribution loop in the city centre area; a project which has been advocated before to provide the city itself with rapid transit rail coverage.

ADDITIONAL DETAIL

Authority: Cleveland Transit System.

Route length: 19·02 miles (30·6 km).

Stations: There are seventeen stations, several of which are equipped with escalators. Platforms of minimum 300 ft length include both island and side type.

Rolling stock: Besides the 20 new cars there are 88 older type cars, approximately 48 ft long and driven by 55 hp motors. They operate singly or as two-car units. Seating capacity is 52 in the older type cars and 80 in the new 70 ft cars.

Signalling: The automatic block system operates with 3-colour light signals and train stops.

Traction current: This is supplied through an overhead wire system at 600v DC.

In 1968 the Shaker lines carried approximately 5¼ million passengers, and the Cleveland Transit line about 16¼ millions. The airport extension will undoubtedly add considerably to the latter figure, since it not only opens up new territory but serves an airport ranking as one of the largest in the country.

NEW YORK

The City of New York consists of the Island of Manhattan, the Bronx (across the Harlem River to the north), the Boroughs of Brooklyn and Queens (on Long Island east of Manhattan, between the East River and the Atlantic), and part of Staten Island (south-east of Manhattan). Metropolitan New York takes in also part of New Jersey on the mainland and additional parts of Long Island. Of the city's eight million inhabitants, 1,700,000 live in Manhattan. Except for parks and other public spaces, Manhattan is completely built-up with avenues that run north–south, and are mostly numbered from First Avenue nearest the East River; and streets that run east–west and are nearly all numbered, upward from south to north. Broadway cuts diagonally across a uniform pattern of rectangular blocks, from Manhattan's south–east tip, past Central Park and then northward all the way to the Bronx.

The author's experience of New York is limited to a few days' sightseeing, and riding about to observe the operation of its subway (underground) system, the busiest in the world. Of this bustling sub-city of travellers, shoppers, noise, snack-bars and subway trains the only impressions relative to this book are of busy in-town subway lines that surprisingly ended up at the seaside (Coney Island), and at the World's Fair (in the summer of 1964); and of generally mingling with the tens of thousands not on holiday who, like the author, spent part of each working day travelling in crowded subway cars. A sizeable proportion of New York's working population flows daily from North Manhattan and the Boroughs, into Lower Manhattan – an area mostly devoted to commerce and finance – and the subway carries the majority of them.

It was principally to cater for the predecessors of these working people that New York's first rapid transit railroads came into being about ninety years ago. Before then, workpeople lived by necessity near enough to their employment, mostly located in Lower Manhattan, to walk or to ride there by horse-bus or tram. Industry moved farther afield, away from Lower Manhattan, when this crowded area literally ran out of space. Then in 1870 the first short elevated railroad was built over the streets of downtown Manhattan, and this was succeeded by other elevated lines, all private enterprises, extending into northern Manhattan, following a shift of population in that direction. Employers returned to Lower Manhattan and a commuter movement began that grew in proportion to New York's expansion. Patronage of the 'El' increased as connecting lines were built in the developing boroughs of Brooklyn and Queens, until at the height of its activity the Elevated system was carrying a million passengers a day – and it remained a principal carrier until the subway assumed that role.

By the end of the nineteenth century Manhattan space had reached a high premium. Incipient skyscrapers were rising from demolition sites of old tenements and the like, and to meet the need for additional non-land-consuming urban rail routes, with frequent stations, an underground railroad was the obvious answer. The first of such, appropriately named the subway, was four years in building, from 1900 to 1904 – a 25 ft wide rectangular tunnel, mostly drilled out of rock and covered over. When opened it ran only from near Brooklyn Bridge, via Grand Central Station and Times Square up to 145th Street; but during its construction approval was given for extensions into the Bronx, and subsequently under the East River to Brooklyn.

The first subway was a municipal project, constructed with funds secured from the sale of city-owned bonds. It was operated by the Interborough Rapid Transit Company (IRT) and became the nucleus of this Company's system of subway lines in Manhattan, the Bronx, Brooklyn and Queens, and privately owned lines in Manhattan and the Bronx. A second group of subway lines was built between the years 1913 – 1920 by the Brooklyn-Manhattan Transit Corporation (BMT), whose system included city-owned subways in Brooklyn, Manhattan and Queens, and privately owned elevated lines in Brooklyn and Queens terminating at the Manhattan ends of East River bridges. A third group of lines, the

Independant system (IND), built between the years 1925 and 1930, consisted of subway lines owned and operated by the City of New York, located in Brooklyn, Manhattan and the Bronx.

Although there was never any question of the efficiency of urban elevated railroads, in modern New York their disadvantages, which included the obstructive effect of their supporting pillars on motor traffic and the din of many trains passing over steel structures, outweighed their advantages. Consequently they were progressively removed as the subway system developed, and today there are only two completely elevated lines, existing as parts of the 3rd Avenue and Myrtle lines.

Continuing with the subway system's later history, in 1940 the IRT and BMT systems, until then privately operated, were acquired by the City of New York and became part of the unified New York City Transit System. In 1953 the present New York City Transit Authority was created to take over rapid transit and surface lines, and since March 1968 this Authority has come under the jurisdiction of the state-created Metropolitan Transportation Authority. Finally, the present maps of the New York subway system no longer bear the somewhat confusing legends. IRT, IND, BMT, as with unification these former definitions of divisions have been abandoned.

The board of the new Authority has wide responsibilities. Briefly, these are concerned with general mass transportation in the New York State sector of the metropolitan region, and thus cover a wide field. The Authority's concerns include the balancing of transportation media in that area; modernization of commuter railroads as well as subways; and closer integration of road, rail and air systems. Under such a comprehensive heading comes provision for such items as high-speed cars for the Long Island Railroad, besides new subway cars, proposed subway expansion projects, and projects that would affect, or would be part of, more than one form of transportation. An instance of the latter is a projected new four-track East River tunnel, which would be shared by subway trains together with Long Island trains routed to a proposed new terminus in East Manhattan. (At present the latter continue to Pennsylvania Station in the west.)

The announcement of proposals for a massive expansion of the existing subway system (eight new routes have been mentioned) which if carried out would involve the outlay of vast sums of public money, illustrates a growing awareness of the importance of adequate public mass transportation in and around New York. A reminder of the city's heavy dependence on the existing system is given in a recent study of the effects of the thirteen-day transit strike in 1966. The estimated losses were considerable and revealing.

The physical aspects of the New York subway system are too many and various for detailed description here. Its network of routes, of whose length more than half is underground and the remainder at surface level or on viaduct, is now the second longest in the world, and the system carries the greatest number of passengers. A distinguishing feature is the quadruple or multi-tracked tunnel which permits express or limited-stop trains and local stopping trains to be operated on parallel tracks on certain stretches. Additionally there are sections with three tracks where the outer ones are for local services and the inner one for express trains in 'up' or 'down' direction depending on the peak traffic flow. At 'local' stations, where express trains do not stop, the platforms are at the side, and at 'express' stations one may change from local trains on outer tracks to express trains on inner tracks, over island platforms: but there are 484 stations on the system and lay-outs vary from the simple side-platform type on outlying viaduct sections, to multi-level or multi-platformed busy interchange stations.

ADDITIONAL DETAIL

Name of Authority: New York City Transit Authority.

Track and route length: Excluding yards and shops there are 726 miles of track on the system. The actual route length totals 239·87 miles (384·9 km) of which approximately 137 miles is underground, 80 miles on elevated structure and 23 miles in open cut, or at surface or embankment level (1968). The track gauge is 4 ft 8½ in (1·435m).

Rolling stock: The Authority owns approximately 7,000 cars. About three-quarters of this total are cars acquired since 1953. A batch of 400 type R40 cars built by the St Louis Car division of General Steel Industries, Inc. bought in 1966 are currently being delivered. A further 400 type R42 cars of new design from the same builders were ordered in 1968. 200 of type R40 and all of type R42 are air-conditioned. All modern subway cars on the system are driven by four 100 hp motors and have dynamic and air braking. Dimensions of the later cars are 60 ft 6 in, length over couplers, and width 10 ft.

Stations: At underground stations the ticket halls are at sub-surface level, approached by stairs from the street. Some stations have escalators or lifts from the ticket hall to the platform level. The deepest station is 191st Street at Broadway, 180 ft deep from track to street level. Of the total of 484 stations, 138 are in Manhattan.

Signalling: The signalling system is automatic block with colour-light signals and train stops. On the Flushing Line an 'indentra' train operation system, similar to the 'programme' operating system on the London system, is used.

Train speeds and service density, etc.: The maximum number of cars forming an 'express' train is eleven, and a 'local' train ten. Express trains average 22 mph (35 km/h) on passenger service, and local trains 20 mph (32 km/h). The maximum service density is thirty-four trains per hour each way.

Tunnel construction: Much of the system is in rectangular tunnel built by the cut and cover method, but there are sections of tube tunnel; under the East River for instance.

Traction current: To meet the requirements of the increased number and higher acceleration characteristics of its cars a large-scale modernization of the Authority's power distribution system has been carried out in recent years. Traction current, at 600v DC through a third rail system, is obtained through centrally-controlled sub-stations, that previously were of the manually-controlled type.

In the statistical year 1967–8 the New York City transit system carried approximately 1,739,000,000 passengers on the whole of its rail and road services. Of that total 1,303,465,841 passengers were carried on the rail system. A basic fare system operates on the subway.

In addition to the NYCTA system, three other rapid transit rail systems operate in the New York area. The first, the Hudson Bay Tube (the PATH system) is referred to in the next section of this book. The other two are: the Newark Subway, a four-mile rail rapid transit system, operating with street-car type vehicles, which acts as a feeder line to the Hudson Bay Tube; and the Staten Island Railroad, operating along the length of that Island with multiple-unit type cars similar to those in use on the former BMT Division of the New York subway system. This 14-mile line, carrying 22,000 passengers a day, is owned by the Baltimore and Ohio Railroad. A proposal has been made for it to be taken over by the City of New York and then modernized.

NEW YORK (PATH)

The tube railway system now known as the Port Authority Trans-Hudson (PATH) transit system began operation in 1908. In that year its newly-opened tube tunnels under the Hudson River provided the first under-water rail link between Manhattan and New Jersey, through which electrified train services carried passengers from Jersey City, and from commuter railroads in developing suburbs on the mainland, directly into Manhattan. The system, operated by the Hudson and Manhattan Railway Company, seemed destined for an increasingly busy and useful life – and this proved to be the case for the next twenty-five years. In the year 1929 its passenger total reached a peak of 100 millions; but after this traffic fell away steadily, for reasons associated with fundamental changes in metropolitan living. Increasing use of private automobiles, the reduction of the working week, decentralization of industry and growth of suburban shopping areas all played a part in the decline of the Hudson and Manhattan Railway Company's revenue, and contributed to the deterioration of its rolling stock and equipment.

In order to arrest this decline and preserve mass transit facilities across the Hudson River, the Port of New York Authority acquired the railroad in 1962, and through its subsidiary, the Port Authority Trans-Hudson, began a programme of re-habilitation and modernization, a task for which PATH was created.

The system basically consists of two rail routes under the Hudson River, through four tunnels of from 15 to 18 ft diameter. The two up-river tunnels form part of the line that runs from Hoboken on the New Jersey side to 33rd Street in mid-Manhattan. The two down-river tunnels form part of the line that now runs between Newark on the mainland and the Hudson Terminal in Lower Manhattan. A branch line in Jersey City links the two tunnel routes.

The new Authority's modernization programme included: the provision of a fleet of new aluminium air-conditioned cars to replace the old cars, some built as early as 1908; modernization of the signalling system and installation of automatic control devices, and an improved power distribution system. Much of this re-habilitation work had been done when on April 30, 1967, PATH began the sole operation of services over 6·3 miles of rail route between Newark and Jersey City, which previous to this date were operated by joint PATH–Pennsylvania Railroad Joint services. The Authority's effective rail route was thus increased from its original 7·9 miles to 14·2 miles (23·1 km).

The most important development affecting the future of the PATH system was probably the implementation of the 'Aldene' plan, which took place on the same date, April 30, 1967. Previous to this, rail passengers from central New Jersey bound for Lower Manhattan detrained from Central Railroad of New Jersey trains at a Jersey Street terminus on the Hudson, from whence they were ferried across the river to Manhattan. Under the new arrangement, CNJ trains were re-routed over Lehigh Valley and Pennsylvania railroad tracks into Newark, enabling passengers to continue into Manhattan either by the improved PATH rail services or by the Penn Central rail services. Following the change-over, there has come the expected increase in passengers using PATH services. In 1961 the total was about 31 millions. In 1965 it had dropped to 26·4 millions, but in 1967, after less than a full year of the new working, it had risen again to 30·5 millions.

Construction work on two more important projects concerning the PATH system is now (1969) under way. The first project is a new Transportation Centre at Journal Square in Jersey City, which will contain a new PATH station, a bus terminal and auto-parking facilities. The second is a new terminal for PATH passengers beneath the World Trade Centre in Lower Manhatten, the Port of New York's new focal point for international commerce. This huge building complex, now under construction, is expected to open in 1970 and to be entirely completed by 1972.

The latest cars on the PATH system comprise 110 driving cars and 52 motored trailer cars, built by the St Louis Car Division of

General Steel Industries, to replace the old original cars, and a further 11 driving and 30 trailer cars, of the same design as the newer cars, for the additional riders created by the 'Aldene' re-routing project. They are 51 ft 3 in long (15·62 m) and each carry about 140 passengers. The system's traction current is distributed by a third rail, at 600v DC.

PHILADELPHIA

Philadelphia, the historical Quaker city on the Delaware River, was founded by William Penn nearly 300 years ago. His effigy stands on top of the fine City Hall, looking over a city that now extends along the Delaware River for about twenty miles and reaches inland for from five to fifteen miles. It was planned originally on a grid iron pattern of streets and avenues which has persisted more or less over all the city's subsequent development. Broad Street and Market Street, running north–south and east–west through the city centre, are the main thoroughfares which formed the base for the original plan.

The metropolitan area of Philadelphia today contains a population of about four millions. Its public transportation needs are met by an extensive system of rail rapid-transit lines, trackless trolleys, subway-surface cars, auto-buses – and suburban commuter rail-roads that once were poorly equipped and losing patronage, but are now in much better physical and financial shape.

The city's urban rapid transit system consists of subway and elevated railroads, and subway-surface tram services routed partly through their own tunnel and partly through tunnel shared with rapid transit trains. The first two categories comprise a group of five lines: the Frankford Elevated, the Broad Street Subway, the Market Street Elevated and Subway, the Ridge Avenue Subway and the Locust Street–Camden line, all operated until September 30, 1968, by the Philadelphia Transportation Company and subsequently by the Southeastern Pennsylvania Transportation Authority (SEPTA), a public agency which acquired the former Company. On February 15, 1969, the Locust Street–Camden line was absorbed into the newly-opened Lindenwald line, which is operated by the Port Authority Transit Corporation, a subsidiary of the Delaware River Port Authority.

The oldest rapid transit line is that along Market Street, about five miles of which was opened in 1907, running partly in subway and partly over elevated structure. The four-track tunnel at its city end was in 1955 extended westward under the Schuylkill River, utilizing *en route* a tunnel under the river built in the depressed 1930s as a 'make-work' job. At 32nd Street, subway and tram routes now divide into their separate tunnels, the latter continuing only a short distance in tunnel before emerging into western Philadelphia as surface routes.

The Frankford Elevated, opened in 1922, makes connection with the Market Street line at 2nd Street near the Delaware River, and 'through' train services are run between the outer terminals of each line, whose track gauge throughout is 5 ft 2¼ in (1·581 m). Philadelphia's first wholly underground line, the Broad Street Subway, came into service in 1928 between City Hall and Olney Avenue, along the northern line of Broad Street. From just south of City Hall, its tunnel is now four-tracked as far as Erie Avenue, permitting express train services between those two points. Erie Station is thus a natural turn-out point for a proposed branch line which will continue north-eastward to the outer suburbs.

The Ridge Avenue Subway opened in 1932 is just a short spur off the downtown end of the Broad Street line, but it serves a shopping area and provides interconnection with the Lindenwald line. The Locust Street–Camden portion of the latter, a subway-surface line first opened in 1936, links the inner city area with the Camden district over the Delaware River, which it crosses by the Benjamin Franklyn Bridge over tracks located outside automobile lanes. The Lindenwald line is a physical 10½ mile surface extension of the latter, in a south-easterly direction. Over its whole 14½ miles, between Philadelphia 16th Street and Lindenwald, there now run high-speed commuter-type trains newly built by the Budd Company. The track gauge of the Lindenwald, the Broad Street and the Ridge Avenue lines is standard 4 ft 8½ in. Interworking between these three lines and the Market Street–Frankford lines is therefore not possible because of the difference in track widths and rolling stock.

North-westward from the 69th Street western terminus of the Market Street line runs another railroad which comes under the descriptive heading of 'rapid transit'. It is officially known as the

Norristown Division of the Philadelphia Suburban Transportation Company, which also operates buses and high-speed street cars in the city's western suburbs. The Norristown line is eighteen miles long, and is operated by an electrified service of trains made up of single or two-car units. There are twenty-one intermediate stations on this self-contained line, whose track is standard-gauge. Its traction current at 600v DC is distributed through a third rail system, and signalling is by the automatic block system with colour-light signals.

The acknowledged excellent commuter services operating in metropolitan Philadelphia have their focal point in the inner city area known locally as 'Center City'. Many downtown underground stations have direct entrances from buildings and shops, and a great complex of underground passageways and concourses in the vicinity of City Hall make it possible for people not only to cross streets in safety, but to walk under cover to no less than eight underground stations, besides the Pennsylvania suburban and the Reading Railroad terminals, and numerous buildings and stores.

An additional facility for this area has been proposed. It will take the form of a new rail subway, to be built at an estimated cost of $48 millions, to connect the two surface railroad stations and allow interworking of each Company's suburban trains over a common linking line. Furthermore, a huge 'Transportation Centre' just north of Market Street has also been proposed, which at its various levels would provide auto-parking, bus stations, pedestrian walkways, sub-surface concourses, and at its lowest level a railroad station. It would in fact extend the existing passageways system into the area of the new Centre.

Additional to the foregoing are planned early extensions to the Broad Street Subway (6·4 miles to the north and 1·2 miles to the south) that are part of a comprehensive mass transit plan on a regional basis, and would be a prelude to much greater expansion of the rapid transit system over the next decade or two.

ADDITIONAL DETAILS

Relating to the South-eastern Pennsylvania Transportation Authority and Delaware River Port Authority rapid transit systems.

On the whole of its subway and elevated rail services together with its road services, the Philadelphia Transportation System in 1968 carried 278,304,000 passengers. Of this total, about 70 million passengers were carried on the subway and elevated services. The maximum number of trains per hour in each direction of the Broad Street Subway is 25, and on the Market Street line 26. The maximum number of cars making up a train is six.

For any distance within the city limits there is a basic fare. This applies to any of the Authority's vehicles, but there is a small charge for first and subsequent changes.

Rolling stock (SEPTA) : On the Frankford–Market line, stainless steel cars manufactured by the Budd Company of Philadelphia are of a type that can be operated singly, and a type that operate as coupled cars. The single cars have controls at each end, whilst the coupled type have only one control position. They are 55 ft long and have four motors per car.

Rolling stock (Delaware River Port Authority) : Stainless steel cars manufactured by the Budd Company operate on the new 16th Street-Lindenwald rapid transit railroad. The cars, each driven by four DC motors, have rapid acceleration characteristics and are capable of attaining 75 mph in as many seconds.

Signalling and power (SEPTA) : An automatic-block system with colour-light signals operates. Traction current at 600v DC is distributed through a third rail system.

Signalling and power (Port Authority Lindenwald line) : An automatic train operation system is a feature of this new line. It accelerates the train at maximum performance up to the limits of allowable speed, regulates power to maintain allowable speed, and then initiates and regulates braking. Train door closing, train starting and train door opening are performed by the train attendant by means of push-button control. Trains can also be signalled through normal track-side signalling, and can be manually operated by the attendant. Traction current at 740v DC is distributed through a third rail system.

SAN FRANCISCO

The largest single rail rapid transit project the world has yet seen is nearing completion in the Bay Area of San Francisco. This particular form of mass transportation is new to the State of California, and with its advent in 1971–2 the people of San Francisco and a score or so of towns in the mainland counties of Alameda and Contra Costa will possess a rail system featuring all that is modern and forward-looking in the field of rapid transit. Nearly everyone in the three counties will be affected in one way or another by the new railroad, but it will principally benefit those living within, or intending to live within the wide 'feeder' belt along its seventy-five miles of route.

The extreme importance of the new railroad to the whole Bay Area, however, lies in the fact that it will provide the first and only direct rail link between the San Francisco peninsula and the mainland, and be the first to cross beneath the Bay in tunnel. The existing links by Bay bridges and ferry cannot adequately handle the peak-hour surges of traffic now, and this volume of traffic tends to increase between the mainland's developing towns and the magnet that is San Francisco. Here for prestige reasons are located the headquarters of the largest commercial enterprises, drawn to the tip of the peninsula in the first place because it was, and is, the shipping gateway to the far East. It is reasonable to suppose that the city's future as the focal point for northern California's business and social activities will be more firmly established by the advent of the new railroad system.

Geographically, most of the Bay Area Rapid Transit District is on the mainland, and much of it lies to the east of the Berkeley Hills which form a barrier between it and the Bay. The railroad, in piercing these hills and continuing into the valley land on the other side, will not so much generate traffic as anticipate it. These inland areas are relatively thinly populated now, and whilst San Francisco

144

Above: Mexico City: pneumatic-tyred rolling stock for the city's
etro; *below*: Buenos Aires: multiple unit stock of Spanish manufacture
service extended Line E.

36. *Above:* Tokyo: morning peak scene at Tokyo, Japanese National Railways; *below:* Osaka: morning rush hour at Umeda Station.

7. *Above:* Tokyo: Monorail between Tokyo International Airport and Hamamatsu terminus served by Daimon station on Subway 1; *below:* Tokyo: prototype aluminium electric cars for the subway for service on Line 9.

38. *Above*: Tokaido line: author with the 'Limited'; *below*: Osaka: train
on Subway Line 4 in double-track tunnel constructed by shield method

itself cannot physically expand to house more people, the inland settlements beyond the Berkeley Hills are expected to in any case, and to double their population by 1980. The new rapid transit system therefore has two main functions: to relieve the presently crowded Bay bridges and ferry services and to serve developing areas on the mainland.

Historically the BARTD (to quote the Authority's title) was created in 1957 to plan and build the system, and within its taxation powers to finance construction by property taxes and other means. In 1963 the first issue of bonds to finance construction was sold, and in 1964 actual construction of the system began. Within the ensuing years, trials of various forms of traction and current collection were carried out over an experimental stretch of track in central Contra Costa. Construction work was begun at many sites along the route, twin tunnels were driven through the Berkeley Hills, and the actual Trans-Bay Tube was started, all within the first year or two. Today (1969) the Bay Tube rests on the sea-bed and is ready for equipping, and most of the major tunnelling work and construction of elevated structure along the route is completed.

Probably the most radical of the new system's features is its wide track gauge. Exhaustive theoretical and practical study of the behaviour of trains travelling at high speed over elevated structure (one-third of the system is elevated) revealed that a track width of 5 ft 6 in offered the greatest stability. The principal factors in this study were the light-weight trains themselves and the effects upon them of lateral air pressures from winds of up to 85 mph, and gusts of higher speed, at an elevation of 30 ft above ground level. Wind speeds of this order are the maximum likely to be experienced in the Bay Area, a region prone to fairly high winds. Light-weight construction of trains was considered essential to permit rapid accelera tion to high speed, a characteristic called for in order to attain maximum speeds up to 80 mph and service speeds of 45 to 50 mph, necessary to maintain high-frequency services over peak periods.

There was no need for the system to conform to the standard 4 ft 8½ in track gauge, as it is a self-contained railroad entity. This is helpful also in respect of the operation of the system's trains, which will be by automatic control. Five different control concepts were tested and the Westinghouse Electric Corporation's system finally accepted. This incorporates three sub-systems, which are respectively train protection (the insurance of safe spacing of trains), automatic train operation (which controls starting, stopping, speed regulation and the operation of train doors), and the continuous monitoring of all train movements from a central computer control point in Oakland on the mainland. In practice the system will work similarly to the automatically controlled lines on systems in other major cities. The duties of a train attendant in the leading car of each train will include supervision of the train's performance, attendance on the passengers' needs, and ultimate control of the train's progress or stopping only in an emergency. The other duties normally performed by motormen and train guards (starting, accelerating, braking and stopping as dictated by line-side signal, and operation of doors) will be entirely automatic on the BART system.

Electric power for the system will be bought from the Pacific Gas and Electric Company and transformed to 1,000v DC for traction current. This high voltage was selected primarily to meet the BART system's high-performance requirements. Fare collection is also to be a largely automatic operation, embracing a system of gates, ticket-vending machines and money-changing machines that will function pneumatically and electronically, employing the same basic principles described in the section of this book dealing with London's Underground, but probably in a more sophisticated way.

Of the many distinct engineering projects forming the whole BART construction complex, several are of considerable magnitude, entailing years of work in themselves. The Berkeley Hills tunnels are an example: 446 working days had already been spent in excavation before the final break-through (in the middle) on February 24, 1967. The twin tunnels, 3½ miles long, 20 ft in diameter and set 100 ft apart, are braced with horseshoe-shaped ribs and lined with concrete 18 in thick. To maintain the line's course it was necessary to drive straight through the Hayward Fault, an active earthquake zone where seismic movement through the ages had tilted, fractured, and in some places ground rock layers to a fine powder. No less than nineteen distinct kinds of strata were encountered.

Of the seventy-five miles of BART route, twenty-five are on aerial or viaduct structure. Almost all the sixteen miles of route between Oakland and Hayward is thus raised, the method generally employed on the system being to cast *in situ* concrete T columns about 20 ft high. These support side-by-side pre-cast and pre-stressed concrete roadbed girders, which are virtually units of permanent way hoisted on pillars. Each unit is massive and heavy, measuring 11 ft 8 in in width, 4 ft in depth and up to 98 ft in length. The biggest weigh nearly 120 tons. They are of a uniform cross-section design which permitted mass production at a girder plant, from which they were conveyed to their sites on specially designed ninety-wheel trailers. (Incidentally, the train ride from Oakland to Hayward could well be one of the most interesting on the whole route, providing elevated views of the Bay, which it closely skirts, and the peninsula beyond.)

In San Francisco and downtown Oakland and Berkeley the lines descend into subway, those under Oakland necessitating three subway tunnels where the route from San Francisco splits into three: northward to Richmond, eastward to Concord and southward to Fremont. Construction methods for these subways included cut and cover and mining, the most complicated tunnelling being that under San Francisco's main thoroughfare, Market Street. Here the BART trains will run in tunnel seventy-five feet below ground. Above them in tunnel will run San Francisco's street-car services, diverted below ground from their present surface tracks; and at sub-surface level will be large passenger-circulation areas. The BART tunnels here have been driven by the shield and machine method: the machine, aptly named a 'mechanical mole' comprising basically a revolving head armed with teeth boring a tunnel eighteen feet in diameter. In this locality the soil is water-saturated, and where it was not possible to draw down the water table by sinking deep wells, mining was done under compressed air conditions.

This virtual disembowelling process under about a mile and three-quarters of Market Street went on mostly out of sight and mind of the many shoppers and people going about their business in this busy thoroughfare – out of sight, that is, except for the huge excavations two blocks long and seventy-five feet deep at station sites. Three of the stations along the street lie practically submerged at fifty–sixty feet below water level, and because of the considerable hydrostatic pressure against their structures, their construction is particularly heavy and massive to counteract a 'floating' tendency. This deep-level stretch of the BART system continues eastward under the floor of the Bay to the mainland. The manner in which it does so, in the Trans-Bay Tube, is described in the introduction to this book. It is impossible to convey in print the tremendous scale and range of this whole BART operation, from the 12,000-ton sections of the actual Tube gently lowered and precisely placed end to end under the Bay, to plastic components for cars small enough to be held in the hand.

The author saw it all begin near Walnut Creek, nine miles west of the Berkeley Hills, in 1964. Arid, sandy dunes were being levelled to put down the very first rails for the Diablo Test Track. There was not a dwelling to be seen westward, and only a few near the engineers' huts back towards the east. Previously the author had been taken to the top of Mount Tamalpais, in Marin County over the Golden Gate Bridge, to see in one vast sweep the whole of the BART terrain. Nature has been abundant and prolific on the Bay slopes and the islands, and man has enhanced its beauty with many great bridges over the San Pablo and San Francisco Bays; and of course San Francisco itself. Add to these conditions a benign climate and one has the reason why more and more people desire to live and work here – and the impelling reason for efficient and comfortable rail transportation that the BART services will provide.

WASHINGTON

As far back as 1909 a prominent citizen of the Nation's capital spoke of the desirability of a subway for Washington – and over succeeding years this theme has been continued through a large number of speeches and writings. Of all the matter printed during this long period – subway plans, graphs, reported speeches and painstaking surveys, much has probably gone to the pulp mills; but the considerable sum of their effort has not been wasted, because the Washington Metro is at last a tangible project, differing only from earlier concepts in that it is geared to contemporary and future development.

On September 8, 1965, a basic twenty-five mile urban rapid transit system was authorized for the District of Columbia. In March 1968 a more comprehensive plan and programme for a regional rapid transit system was adopted, and revised in February 1969, which will expand the basic system into a 97·7 mile (157·2 km) network serving parts of suburban Maryland and Northern Virginia, as well as the District of Columbia. The proposed Regional network would comprise 29·9 miles (48·1 km) of route in Maryland, 30·1 miles (49·4 km) in Northern Virginia and 37·7 miles (60·6 km) in the District of Columbia.

A start on actual construction of the system began early in 1970, the participating public having voted to give local financial support to the project. Five of the suburban jurisdictions in the National Capital region indicated to Congress their strong desire for rail rapid transit by a more than 70 per cent approving referenda to authorize the issuance of bonds for construction, to encourage massive Federal aid in the shape of local grants. All participating areas have their representatives in the Washington Metropolitan Area Transportation Authority, a controlling body created to deal with the expanded project, having taken over the functions and duties of the smaller National Capitol Transportation Agency in 1967.

In geographical plan, the Washington Metropolitan Transit area roughly resembles a fully foliaged tree, with the Potomac River forming its trunk, and the Washington Channel and Anacostia River its two other limbs. The business area is at the centre and the Federal Government establishment complex just to the south-west. (The Government is the largest employing agency in Washington.) Until the 1950s the city and its immediate surrounds managed its transport reasonably adequately with suburban railways, tramcars, buses and private cars. (In 1962 the tram system was discontinued after 100 years' service.) It was seen to be inevitable that as the city grew, the continuous expansion of its highway system would never be sufficient to accommodate all the surface transport, especially in the downtown area: and that for future years, with population increasing from 2·6 millions in 1967 to anticipated totals of 3·5 millions in 1980 and 4·2 millions by 1990, an underground railway system must be built. Moreover, employment growth showed a trend to outward movement, and a radial expansion of the basic underground railway system was considered prudent to serve these new employment areas.

For these reasons the authorized system and the proposed regional network have been planned to meet expected intensive commuter demands, with services operating on a frequency basis of a train every two minutes on its busiest lines (the Rockville and Glenmont sections of the U-shaped line which will run north-west and north from the city centre, and the Greenbelt Road line to the north-east) during morning and evening periods, and a train every four minutes during peak periods on other lines.

In the north, rapid transit trains will run partly on stretches of the Baltimore and Ohio Railroad right of way, and in the south-west partly on the Pennsylvania Railroad right of way. Otherwise, the lines will have their own exclusive rights of way. They will be at deep level or sub-surface level generally in inner areas and at surface level or on viaduct elsewhere. About thirteen miles of the basic system will be in subway, in shield-driven tunnel approximately 100 ft below ground at its lowest level, and in tunnel constructed by the cut and cover method where shallow levels permit.

Stations will vary according to their depth, and the medium or locality in which they are situated – in rock for example they will be dome-shaped with fly-over bridges leading from mezzanines extending over both tracks. (Tracks will be of standard, 4 ft 8½ in gauge.) The platform length of 600 ft will be sufficient to take trains of eight cars. There will be 86 stations, of which 53 will be underground.

The cars, a prototype of which has been on view for some time, are 75 ft long, 10 ft 10 in high and 10 ft wide. They will be coupled in two-car units with an operator's cab at each end of the unit. The operators will attend the train and passengers, giving information generally or specifically as it is received over the train's radio from Control. The train's progress will be governed by automatic train control, which the operator can over-ride by manual control in emergency.

Initially, services are expected to begin in 1972 on six miles of the in town section of the U-shaped line terminating at Rockville and Glenmont respectively (that portion which runs between Dupont Circle and Rhode Island Avenue via the B and O Union Station). The whole network is expected to take twelve years to complete. Its estimated cost of $2·5 billions (£1,000 millions approximately in the U.K.) is expected to be met to about one-third from revenue (excluding bond interest) and the remainder from Federal Government and local jurisdiction sources.

MEXICO

MEXICO CITY

The 1968 Olympic Games held in Mexico City were more to that city than just successful athletic events. They focused world attention on Mexico and gave a fillip to that country's already considerable tourist traffic, attracted by Mexico's varied and unique range of archeological remains and picturesque scenery. Mexico City, in particular, is affected by this activity in so far as an expansive programme of hotel construction was undertaken to accommodate the growing tourist traffic and the influx of football enthusiasts for the 1970 World Cup.

The city's remarkable growth in recent years may thus be attributable partly to the tourist and visitor influx and partly to the country's considerable general economic development over the last two or three decades. The city's physical growth is illustrated by its population total, which stood at 1,448,000 in 1940, in 1967 had reached 6,300,000, and with an annual growth rate of 6 per cent is expected to reach 7½ millions by 1970 (9½ millions in the Federal District).

Figuring largely in Mexico's development is its investment in construction, in public and private works, of which the new Metro system for Mexico City is an example. This is being constructed, under the authority of El Servico De Transporte Colectivo, to alleviate surface traffic congestion, and to meet the needs of this large city for an efficient and speedy means of mass public transport, operating in its own right of way – in this case, underground.

The first stage construction of Line 1 began in June 1967, and a year later work began on Lines 2 and 3, which together with Line 1 will form the initial three-line Metro network for Mexico City. Line 1 comprises a double-track, sub-surface line with a route length of 7·8 miles (12·6 km) and sixteen stations. During development of this line it was decided, in view of considered transportation needs of a densely populated area near to the proposed western

152

terminus, to extend Line 1 under Avenue Jelisco, thus adding two more stations and a further 1·5 miles (2·4 km) of route to the line, all of which is underground.

It runs directly beneath centrally situated east–west main thoroughfares, between Tacubaya station in the west and Zaragoza station in the east. A short northward spur line from the vicinity of the eastern terminus ascends to surface level and gives access to the line's car storage and maintenance depot. The rectangular, concrete running tunnels, at shallow sub-surface level, were excavated from the surface.

The Metro will be the second system in North America, after Montreal, using pneumatic-tyred rolling stock, similar to that on the Paris Metro (described elsewhere in this book). French interest in the project, besides consultancy, takes the form of a French fifteen-year loan to help finance construction. On Line 1 station platforms are in general approximately 503 ft (153·4 m) long, (although some are considerably longer) and are sufficient to accommodate trains of up to nine cars. They will be equipped with shallow escalators where passageways descend beneath the running tracks to connect with both side platforms. The composition of nine-car trains will be two end motored-driving cars, four motored non-driving cars and three trailer cars.

Initally, trains will operate at up to four-minute intervals, running at a service speed of 32 km/h including stops. Each train will have a capacity of 1,500 passengers, affording, at this train frequency, a line capacity of 22,500 passengers an hour in each direction. The journey time between Chapultepec and Zaragoza stations on Line 1, 23 minutes, will compare favourably with the same journey by automobile, which takes about 45 minutes.

The route of Line 2 is northward from Pino Square station and then westward, crossing the route of Line 3, which runs north-south and connects with Line 1 at Balderas station. The operational date for Line 1 is expected to be about June 1969, and that for Lines 2 and 3 about November 1970.*

* The initial stretch of Line 1 opened for traffic on September 4, 1969, and is now carrying 25,000 passengers daily.

ARGENTINE

BUENOS AIRES

More than seven million people today live in Metropolitan Buenos Aires and its suburbs, an area of about 80 square miles resembling in plan a vast chequer-board of squares and rectangles formed by the geometrical pattern of its streets. The city, which was founded about 400 years ago, grew most rapidly in size and importance towards the end of last century, its population increasing fourfold in some twenty-eight years. Railways, refrigeration and shipping were important factors in this growth: the railways focusing all their major routes on to the capital and bringing the products of the hinterland in bulk to the port; then the development of refrigeration which stimulated trade in meat exports – and consequently stimulated development of the port's facilities to accommodate increased shipping; all of which contributed to the transformation of a small port into the biggest in South America, and one of the biggest ports in the world.

Buenos Aires, the largest city in the southern hemisphere, is the main commercial and industrial centre for the Argentine Republic, and is well provided with theatres, hotels, restaurants and similar amenities for its own population and for its many visitors. Its public urban transportation is represented by efficient bus, trolley-bus, subway and taxi services.

Compared with those of other major cities of similar size, the Buenos Aires subway (underground railway) system is small, but is considered for enlargement in the future. The city's first subway (Line A on the map) is more than fifty years old. It was opened in stages between December 1913 and July 1914 and operated by the Anglo-Argentine Tramways Company. It runs between Primera Junta in the west and Plaza Mayo in the dock area. Line B, roughly in parallel with Line A, was opened in 1930 and runs between Leandro N. Alen at the docks and Frederico Lacroze to the north-west, being first operated by the Buenos Aires Central Terminal

Railway. Lines C, D and E (Retiro–Plaza Constitution, Florida–Palermo, and Bolivar Street–Av La Plata), were opened between 1933 and 1944 and operated by a Spanish Company.

All five lines from 1936 to 1952 were operated under the control of the Buenos Aires Transport Corporation, formed to co-ordinate the city's passenger transport, whose authority passed to a department of the Ministry of Transport in the latter year. In 1963 the 'underground' system came under the control of the Subterraneos de Buenos Aires, which is now responsible for the running of this part of the city's public transport.

The complete physical integration of all the lines into one system is not possible, although they all have standard gauge (4 ft 8½ in) track. The traction current for Line A is at 1100v DC, collected through an overhead wire system; that for Line B is at 550–600v DC via a third rail; and on Lines C, D and E the traction current is at 1500v DC through overhead wires. On these last three lines the cars are, however, interchangeable. The trains on all lines are made up of multiple-unit stock, that on Line A being of British manufacture, on Line B of American manufacture, and on Lines C, D and E of German and local manufacture. The latest cars principally for servicing the extended Line E are of Spanish manufacture, the car bodies being locally built.

Similar to the London Underground system in respect of its composition of formerly separate lines built over a long period, the Buenos Aires system's constructional details vary considerably. The tunnels on Line A, for instance, are rectangular and were built by the cut and cover method. They are 25 ft 3 in (7·7 m) wide and have no centre supports. Those on Line B are partly rectangular with centre support, are 27 ft 9 in (8·45 m) wide and were built by mining methods; whilst those on Lines C, D and E include horse-shoe shaped tunnels and tunnels with arched roofs, differing slightly in width from the former lines. Stations in some cases, although on different subway lines, are on the same street. In these cases, although the stations are some distance apart from each other, they bear the same names. With the exception of Line E, all lines have stations providing interconnection with the Argentine surface railways.

The subway lines do not at present extend far (from the downtown area) into the city's built-up districts, although two proposed extensions would penetrate farther into the north-western districts as surface extensions. These would be prolongations of Lines B and D from F. Lacroze and Palermo stations respectively, and in this connection it has been reported that a consortium of consulting engineers have offered to draw up plans for a 3·7 km extension of Line D with five stations. Recent short extensions to Line E at its western end (to La Plata) and near the dock area to Bolivar Street, brought the total length of route on the system to 19·6 miles (31·66 km) and the number of operative stations to 56. A new downtown line is also proposed to parallel Line C and provide interconnection with all five of the present lines. It would run in a north–south direction from Sante Fe to Constitution and be routed partly through a rehabilitated stretch of railway tunnel, at present obsolete.

In 1969 the Buenos Aires subway system carried approximately 340 million passengers.

JAPAN

TOKYO

The reason for the immense drive behind Japan's subway and rapid transit construction in recent years lies in that country's economic expansion, its effect in concentrating people into large cities, and the need to provide these people with adequate mass transportation. Tokyo and Osaka are the cities most affected, the trend continuing towards even larger concentrations into urban belts surrounding them. Recent reports about the region known as Greater Tokyo, already the largest metropolis in the world, quote it as adding over three million people every five years to its present population total of eighteen millions.

The city proper is undoubtedly changing rapidly. When the author was there in 1964 he observed the great number of traditionally small two-storey houses, in a maze of little streets (mostly un-named) that increasingly were becoming short cuts for traffic snarled in main road traffic jams. Now, in 1969, there are urban motorways cutting across the old maze of streets, and the little houses are giving way to tall apartment and commercial buildings: but the old commuter movement of fantastic proportion is apparently as great as ever.

No city in the world matches up to Tokyo in the sheer amount of commuter travel, or in the number and density of its rail commuter services. No less than six electrified main lines and nine private rail lines serve the capital, several of the latter having their terminals on the Japanese National Railway's Yamate Loop Line. The situation here a few years ago could truthfully have been described as climacteric. These private lines provided no through services into the city, with the result that commuters were largely channelled on to the Loop, where they crammed the trains and debouched in their tens of thousands from two or three principal downtown stations. The lines were carrying up to 300 per cent of their normal passenger-load efficiency totals in a situation that was admittedly

157

aggravated during winter months, when the passengers' overcoats and thick clothing added just that much to loading and unloading times as to be noticeable.

Some idea of commuter rail travel then is conveyed in the passenger totals using the Ginza side of Tokyo Station, the busiest side that opens on to the commercial and shopping areas. It was estimated that 500,000 people passed in and out of these portals within a 24-hour workday period. (One of the busiest railway commuter stations in London, Waterloo, handles an estimated 156,000 passengers daily).

It was to relieve this situation that new and existing subway lines were planned to make physical connection with surface commuter railways, and by interworking of both services over each other's tracks to enable passengers to reach the heart of the city, and vice-versa, by surface or subway, without a change of train; the surface lines benefiting of course by a relative relief of pressure on their services.

In March 1965 the first through operation between the Tobu and Toyoko lines and the new No 2 subway, the north–south Hibiya line, began. This has since been followed by the connection of private railways to both ends of the No 1 subway, which at present runs between Oshiage in the north and Daimon in the south but is presently being extended 12½ miles (20 km) farther south to Nishimagome. The No 5 Tozai Subway, which will take an entirely new route for 19 miles (31 km) from Tokyo's western suburbs, through the city centre to the eastern suburbs, and is now half completed, will permit through working of the JNR Chuo line trains into the city.

There was only one subway line in Tokyo before the war, the No 3 Ginza line, which ran for nine miles between Asakusa in the north, the city centre, and Shibuya in the south. The war, which stopped all work on projected lines and extensions, hammered the city and its railways to such effect that prolonged rehabilitation was necessary, and the original master plan for an extensive subway system was not revived until 1951, when construction work was resumed.

In 1955 the Ministry of Transportation established the City

Transport Council to study an overall plan for improving urban and suburban transport in Japan's three major cities, the outcome of which was, among other things, a recommendation for a network of subways for Tokyo, ten in all, totalling 144 route miles. With characteristic energy this large-scale undertaking was begun and developed to a peak, until work was going on simultaneously at many sites on several lines (in 1964, actually on four lines at the same time). In capital expenditure the figures alone are impressive: for the year 1968 about £61 millions, and from 1962 to 1969 inclusive an estimated grand total for work on nine lines of approximately £349 millions.

The result of all this effort and expenditure, if it has followed the pattern observed in 1964, is an extremely efficient rapid transit rail system, operating for a co-operative travelling public with train services noted for their punctuality, and with well-kept and maintained subway trains. Stemming from continuous inductive-type automatic train control in use on the Hibya and Tozai lines, an automatic train operation device was developed and has operated on the Hibya line since 1962, and is being extended to other lines on the system. In this connection it may be of interest to say that the author was privileged to 'operate' such a train (one not in passenger service) under Tokyo in the summer of 1964. There was about the same smoothness in acceleration and deceleration, and the same degree of accuracy in automatic stopping at marked spots in stations (all most remarkable then), as was later observed on automatically operated London Transport trains.

OSAKA

Japan's second largest city, Osaka, lies beyond the south-western end of the narrow coastal belt, between mountains and shore, that extends along the eastern coast of Honshu, Japan's mainland: Tokyo is 320 miles to the north-east. The 'corridor' between the two is intensely developed and more road and rail traffic passes along it than along any other traffic corridor in the country. It was to relieve the old 3 ft 6 in gauge Tokaido line that the standard-gauge New Tokaido line, the Super-Express line, was built, and ultimately opened in 1964. Its advent increased the importance of Osaka; which was already experiencing a population explosion, as illustrated by the fact that the city, presently with about 3½ million inhabitants, increased its population 21 per cent between the years 1955 and 1960.

There are several reasons why Osaka, although in some respects a smaller replica of Tokyo, has suffered more heavily from over-loading of its public transport. One is the extreme density per square kilometere of its population, and another the relatively small amount of space in the whole city area occupied by its roads. Osaka grew principally as a manufacturing and commercial city, retaining in the process much of its old, narrow street pattern. As motor traffic grew in volume, the city's street cars, trolley-buses and buses became increasingly entangled in slow-moving traffic queues. Their operative speeds fell and their efficiency was lessened when it was most needed, to help move the great tide of commuters into and out of the city in the mornings and evenings. (Osaka's day-time population is three times as great as its night-time population.)

The City Transport Council in 1963 had little alternative but to recommend to the Ministry of Transportation a plan for building six new subways, and extending the pre-war Umeda–Abiko sub-way; the only one (apart from a short unconnected stretch of new line between Port Osaka and Bentencho) that the city then possessed. The first part of the planned subway expansion was put into effect in 1964 with the opening of a northerly extension of the Umeda (No 1) line to Shin Osaka, to provide a subway station adjacent to the newly-opened terminus of the New Tokaido line from Tokyo.

The proposed network of subways will have a total of 71 miles (114·2 km) of route, and is scheduled for completion in 1975. Part of the expansion scheme made provision not only for increased rapid transit (subway) coverage by 1970, when the World Exposi-tion was held just north of the city, but also provides for subway trains to operate over existing privately-operated lines to the Exposition site.

Similarly to Tokyo, the Osaka area has nine private railways in

39. *Above :* Osaka : train on Subway Line 5 equipped with ATC and CTC
apparatus; *below :* Nagoya : Subway tunnel constructed by shield method.

40. *Above:* Nagoya: Sakae Station, platform layout; *below:* Sydney mixed suburban train showing recently introduced double-deck cars.

addition to the Japanese National Railways' lines, but more fortunately five of these have their terminals in downtown Osaka and already play an important role in commuter movement. There are further plans for the extension of surface commuter lines into the city centre, where they will provide interconnection with the expanded subway system. It is most probable that the street congestion observed by the author in Osaka in 1964 has since been considerably alleviated by the implementation, so far as it has taken place, of the Master Transportation Plan.

NAGOYA

At the end of Japan's Pacific coastal belt, 230 miles west of Tokyo and 80 miles east of Osaka, is Nagoya, Japan's third major city. It lies within the favoured coastal region which is attracting population from the hinterland year by year, and is the principal stop for super-express trains on the New Tokaido line running between Tokyo and Osaka. The railways and new trunk roads have added to the city's importance, but independent of these, Nagoya, at the centre of the country's principal industrial zone, has expanded its range of heavy and light industries and added to its population in recent years.

At present there are approximately two million people in Metropolitan Nagoya, but by 1985 this figure is expected to have increased to three and a half millions. In the last war, as one of the largest centres of the munitions industry, Nagoya suffered severe war damage. Then after a decade of re-habilitation the city suffered extensive property damage again, as well as disastrous flooding, when hit by Typhoon Vera in 1959. From this too it has recovered, and resumed the drastic city replanning commenced after the last war. Its streets are wider, and now occupy more than twice the urban area than do Osaka streets of its city area.

Even so, Nagoya's public surface transportation (buses and street cars) was becoming less efficient in areas affected by congestion caused by increasing motor traffic. It was to alleviate this situation, and provide for future economic growth, that five

ι

rapid transit subway lines were planned, starting principally from Sakaemachi, the centre of Nagoya's downtown area. In 1954 construction work began on the first 3·7 miles (6·0 km) of No 1 subway line between Nagoya Station and Ikeshita, which was completed in June 1960.

The line has since been extended eastward, and Nagoya's No 2 line, crossing No 1 at Sakaemachi, also has been opened and extended southward towards its ultimate terminal station at Nagoya Port on Ise Bay. It is anticipated that auto-buses will eventually supplant the city's present street-car services and act as distributors to the subway lines, (as they do extensively now at Sakaemachi). For surface rail transport within the city area there are three JNR trunk railways and the Nagoya Electric and Kinki Nippon private railways.

ADDITIONAL DETAIL

TOKYO

At present (1969), there are five subway lines. Nos 2, 3, 4 and 5 are operated by the Teito Rapid Transit Authority, and No 1 by the Transportation Bureau of the Tokyo Metropolitan Government.

Total planned route length (of the Teito lines is as follows:)
Hibiya (No 2) 20·5 km.
Ginza (No 3) 26·0 km.
Marunouchi (No 4) 35·5 km.
Tozai (No 5) 31·0 km.
Future lines Nos. 6, 7, 8 and 9 will bring the total network's planned route length, including that of Line No 1 (20·0 km) to 234·0 km.
Track Gauge: That of the Hibiya and Tozai lines is 3 ft 6 in (1·067 m), conforming with the two private railways with which they interwork services. All the other subway lines at present operating are 4 ft 8½ in track gauge.
Traction current and distribution: The Hibiya and Tozai lines, and

the Metropolitan Government No 1 line, operate with 1500v DC traction current collected by overhead wire. The Ginza and Marunouchi lines operate with 600v DC current collected through a third rail. The installations conform with those of the surface railways interworking over the subway system.

Signalling: Generally automatic block with three-aspect light signals, but station-time-element and grade-time-element signals (at steep downward grades) also are installed. As mentioned earlier, automatic train operation on Lines 2 and 5 is being extended to other parts of the system. Basically it employs automatic train control apparatus, which takes precedence and ensures the train's safety, and ATO apparatus which operates starting, acceleration, notching off, coasting, speed control and stopping at a predetermined point. The train operator, by pressing a button in the cab, sets the train in motion and its progress to stop is henceforth automatic.

Stations: On the whole system at the time of writing there were more than ninety-two. There is a variety of platform arrangement, but constant features in the most modern stations are good lighting, reflecting coloured surfaces, and ample below-ground circulating space to facilitate movement. The most extensive underground complex is at Grand Ginza, which serves what is probably the busiest shopping district in the world. Here, seven stations of six underground lines, including one projected, are linked below-ground. During construction, the surface area extending over the whole width of Ginza, carrying much traffic including street-cars, was covered with wooden planking able to withstand the weight of fully-laden twenty-ton trucks.

Rolling stock: In 1968 there were more than 1000 cars available on the system. The latest on Line No 5 have car bodies 19·5 m long and are coupled in sets of three. They form trains of six cars, but trains of up to seven-car length operate on the system.

Passenger totals: The rapid build-up of the system and its patronage is illustrated by the daily passenger totals recorded in 1956 and in 1966 respectively. On the four lines of the Teito Authority's system only, the figures were 450,000 and 2,019,000 respectively. In the year 1968 the Teito Authority's four lines carried 909,577,128 passengers.

OSAKA

Authority : Osaka Municipal Transportation Bureau.

Route length of existing subway system : The four lines traversing and radiating from the city centre had in 1968 a total route length of 24·2 miles, but planned extension and new line construction, adding some 17 additional miles of subway route were scheduled for completion in a rush programme up to 1970, to meet extra travel facilities for the World Exposition.

Subway line and station detail : The network so far completed is mainly in tunnel, generally of rectangular section, double-tracked and built by the cut and cover method. Some in-town stretches are in circular shield-driven tunnel. Stations vary in platform length between 120 m and 180 m (396 and 590 ft); the latter, on the principal north–south Line No 1, being of sufficient length to take trains of ten cars.

Signalling : Additional to the basic automatic-block system with colour-light signals, on Lines 2 and 4 there is continuous automatic train control operating under Central Traffic Control, and automatic train operation (ATO) is being developed for the new No 5 line.

Rolling stock : In 1969 there were 438 cars available. The latest type 7000 and 8000 are of stainless steel construction and are equipped with inductive high-frequency radio to enable Traffic Control to communicate with drivers of moving trains.

Track gauge : Standard 4 ft 8½ in (1·435 m).

Traction current : 750v DC distributed via a third rail system.

Yearly total of passengers carried : This total naturally increases with every newly-opened stretch of line. An example is the increase from the 1964 total of 374 millions to the 1968 total of approximately 450 millions.

NAGOYA

The subway system is owned and operated by the City of Nagoya under the Authority 'Nagoya Municipality'.

Route length: The operational lines (Nos 1, 2 and 3) had a total operational route length of approximately 10 miles in 1969. Under construction were east–west extensions to Line 1, north–south extensions to Line 2, and north–west and south–east extensions to Line 3. Under construction also was the subway loop line (No 4) which is to serve the eastern area of Nagoya.

Tunnel and station construction: Much of the existing network is in rectangular double-tracked tunnel built by the cut and cover method beneath wide streets. It was necessary to construct a short section of No 1 line in circular shield-driven tunnel, and some circular tunnel will form part of future construction. In station construction, extensive sub-surface concourses feature, some with shops flanking circulating areas.

Signalling and Traction current: An automatic-block system with colour-light signals and speed-control equipment operates. On Line 2 there is a system of cab-signalling and automatic train control. Traction current at 600v DC is collected through a third rail system.

Rolling stock: The stock comprises cars which are motored on all axles, have driving cabs in leading cars and no cabs in intermediate cars, and are coupled to form trains of three or six cars. To assist in noise absorption the wheel surfaces have rubber sheeting application. The livery is yellow, with purple distinguishing bands on Line 2.

Track Gauge: Standard 4 ft 8½ in (1·435 m).

AUSTRALIA

SYDNEY

In the hospitable coastal belt stretching a hundred miles north and fifty miles south of Sydney live some 65 per cent of New South Wales' population of 4·5 millions, or three-quarters of a million more than in the whole of Western Australia. According to report, Sydney itself has been experiencing a building boom for the last thirteen years, and land for housing is fetching high prices. The length of commuter journeys into town grows as more people move out to the suburbs.

To cater for this growing traffic a rapid transit railway is being built, approximately 8¼ miles long, from central Sydney to the University of New South Wales and Kingsford. It follows a semi-circular route inland from the coast and takes in most of the built-up suburban area to the south-west of the city. Except for a short section of single-track line connecting the new line to the existing metropolitan rail network, all the line will be double-tracked and nearly all of it is being built underground. Construction work, which was already in progress in 1967, is expected to occupy a total period of ten years. The schedule for the opening of the line in sections is planned as follows: Sydney Central to Edgecliff, 1973; to Bondi Junction, 1974; to Randwick, 1976, and to the terminal at Kingsford by 1977.

The new line, to be known as the Eastern Suburbs Railway, will be served initially by trains of four double-decked motorized cars drawing their power from a 1500v DC overhead wire system. A circular suburban rail route between Central Station and Circular Quay, most of which is underground, including four of its six stations, has been in operation for some years. Both this line and the Eastern Suburbs Railway are part of the New South Wales' Government Railways system, which is of standard 4 ft 8½ in gauge.

Additional detail of the existing underground loop line:

The first section of this line from Sydney Central station northward to St James' station was opened in 1926 as an extension of the NSW Government suburban railways service. The second section from Central station northward to Wynyard station was opened in February, 1932, and a month later the main North Sydney rail routes were connected with the main western and southern routes with the opening of the Sydney Harbour Bridge. A connecting link via Circular Quay, opened in 1956, permitted the present loop working. The whole loop comprises about three miles of route.

Rolling stock : Trains can be made up of as many as four motored cars and four double-decked trailer cars, with seating for a total of 764 passengers.

Signalling and Power : The electrified line, double-tracked throughout, is worked by an automatic block system, with trainstops. A five-aspect colour-light system of signalling imposes a speed control according to the colour signals, or combination of colour-signals displayed.

PART III

RAPID TRANSIT SYSTEMS PLANNED OR PROPOSED

Many cities throughout the world either have plans drawn up for intended rapid transit systems or are considering producing such plans. The list alters year by year, as consideration may turn to intent, which in turn may be followed by actual construction. A list compiled for such a book as this can therefore quote only from reports reaching the author up to approximately the date of compilation of the list, and the one that follows does no more than provide a guide to reported activity in the field of rapid transit at that time.

CENTRAL AMERICA

VENEZUELA

CARACAS: This city of 1·8 million people is reported to be awaiting Venezuelan Congressional approval for a 12½-mile Metro system to be built at a cost of between $266 m. and $333 m. Caracas is built along a narrow valley 2½ miles wide and 15 miles long. Mountainous surrounding country prevents normal radial expansion and there is a consequent concentration of traffic along the length of the city, leading to prolonged traffic jams. The new line, which would run between Petare in the east and Catia in the west, would be 91 per cent. in tunnel and 9 per cent. at surface or elevated level. In the rail transportation plan for the city, three additional rapid transit lines are envisaged.

NORTH AMERICA

CANADA

CALGARY: The city has been recommended, as a result of a two-year study to develop, as a long-term project, a twenty-mile railway system.

EDMONTON: The city council has authorized an initial $5m for the study of the practicability of a rapid transit system.

USA

ATLANTA: The Metropolitan Atlanta (Georgia) Rapid Transit Authority, created in 1966, have plans prepared for a rapid transit system to serve the city and four surrounding counties. It comprises a basic network of 40·3 miles of route and possible further extensions. In late 1968 a proposal to finance construction was submitted to the electorate but failed, as only 45 per cent. voted for, and adoption required 50 per cent. In March 1969, a report quoted five cities, Atlanta, Dallas, Denver, Seattle and Pittsburgh as having been selected by the Transportation Department of the US Government to participate in a $1,461,000 programme to design and implement urban mass transit systems.

ST LOUIS: A grant has recently been made by the US Department of Transportation to asist in continued planning for a St Louis regional mass transit system.

BALTIMORE: Based on studies costing $1,510,000, a master plan for a $1·7 billion rapid transit system of seventy-one miles for the Baltimore area has been proposed.

LOS ANGELES: The Southern California Rapid Transit District organization's recent report gave detailed plans for a 'five-corridor' rapid transit network totalling 89·1 miles of route expected to cost $2·5 billions. This was the latest development in plans put forward on a number of occasions in the past. As referred to in Part I of this book, public voting for the project, although considerable, was insufficient for the necessary majority approval.

SOUTH AMERICA

BRAZIL

SAO PAULO: Construction contracts are reported to have been let for a two-line rail rapid transit system for the city to be built at a cost of $38m. (See also Part I of this book.)

CHILE

SANTIAGO: Preliminary work begun in 1968 has now been followed by actual first-stage construction of the initial line of an eventual five-line Metro system for the Chilean capital. By 1973 it is hoped to have this 12·7 km line operational. The whole system will have approximately 57·5 km of route, operating over standard gauge (1·435 m) track, mostly underground.

ASIA

JAPAN

YOKOHAMA: Detailed plans for a four-line conventional rail rapid transit system have been announced, and construction has begun on Yokohama's Lines 1 and 2, The initial 6·8 miles of these two routes are expected to be in operation by the end of 1971. The whole project, which will cost Y155,000 millions (approx. £180 millions) and total 46·3 miles of route, is expected to be completed by 1985. The decision to build such an extensive system was reached after calculation that commuter journeys in Yokohama's suburbs, which lack adequate railway facilities, would have doubled their 1960 totals by 1975, and the existing bus services could not adequately be intensified to meet commuter demands.

HONG KONG

The colony of Hong Kong has vastly expanded its industry and increased its population in recent years. Development north of Kowloon and east of Kwun Tong has intensified the demands on public transport, which accounts for 75 per cent. of all transportation in the colony. The present population total of 3½ millions is expected to have nearly doubled by 1986 and the number of journeys made by public transport is expected to increase in similar ratio. Future augmentation of existing surface transport (buses, trams and ferry) by a system of mass transport is therefore considered desirable. A survey requested by the Hong Kong Government recommends a conventional rail rapid transit system, ultimately of four lines totalling forty route miles, most of which would be underground. The trains would be made up of eight cars, affording a line capacity, with trains operating at two-minute intervals, of 90,000 passengers per hour. The estimated cost of the whole project, $3,404 m (16 Hong Kong $ = £1 sterling), would be spread over a period of seventeen years. Financing would probably be from bond issues plus Government grants.

AUSTRALASIA

Although in New Zealand the cities of *Auckland* and *Wellington* are reported to be considering some form of underground railway construction, the only detailed planning to hand is that relating to the Victorian capital, *Melbourne*, Australia. Awaiting the necessary available finance is an underground project for downtown Melbourne that would comprise four independent but parallel tunnel tracks, encircling the business district and connecting the four major groups of suburban railways serving the city through the Victorian Railways' stations at Spencer Street and Flinders Street.

Depending on the prevailing peak traffic flow, trains on the underground loop would be enabled, through a system of reverse signalling, to operate in the prevailing peak flow directions.

It is reported that work on the project may begin during the financial year 1970–1.

EUROPE

In addition to the European cities whose underground systems appear in Part II of this book, there are at least half a dozen cities where intensive use is made of existing tramways, but as their tracked vehicles become increasingly mixed with other surface traffic in city centres, the authorities have decided to re-route the trams through tunnels built solely for tram traffic – and later maybe, for conversion to take conventional Metro trains. Of the several German cities thus affected, two examples are quoted here:

GERMANY

COLOGNE: is in the process of constructing a 4½ mile (7·5 km) network of tram tunnel. The network consists essentially of a main tunnel line centring on Cologne Central Station, with spurs off to the north and south. Of the city's 200 km of tram route it is proposed to divert some 30 km (15 per cent.) into tunnel. The first stage, 1·4 km of tunnel line, is already operational.

STUTTGART: The first stage of a similar project will provide the city with 1½ miles (2·5 km) of tram tunnel. This will be followed by construction of additional tunnel lines to complete a network 10·6 miles (17 km) in total route length. The double-track tunnels, like those of Cologne, are rectangular, 7·5 m wide, and are being built by the cut and cover method. Stuttgart is to have in addition a V Bahn (connecting railway), which will connect the State railways terminating at Central Station to the State railway system southwest of the city centre.

SWEDEN

GOTHENBURG is considering a plan for a forty-four mile rapid transit rail system on which initial construction might begin in three to four years' time.

FINLAND

HELSINKI, capital of Finland, is one of the latest cities to include a proposal for a rapid transit rail system in its comprehensive public transportation plan. The city and its metropolitan area, with a population of 700,000, is spread over an extensive land area intersected by waterways. The proposed rapid transit system is therefore intended in the first stage only to follow two main traffic routes, although subsequently it may be extended by radial spurs to serve most of Greater Helsinki. The system would be laid to the same 5 ft track gauge as its surface railway system. About one-quarter of the first-stage route (9·0 km) would be underground in double-tracked tunnel. A further two stages of construction, the whole spread over a period of ten years, would bring the rapid transit route to a total of 75 km. In 1970 it was announced that construction work would begin during that year.

USSR

MINSK, capital of Byelorussia, with a growing population that is at present more than 800,000, will be the eighth Russian city to possess a Metro system. The implementation of plans for a twenty-mile Metro network will help to solve a transportation problem and facilitate a larger scale of housing construction.

ITALY

VENICE: A rapid transit rail system has been advocated to connect Venice with the mainland. This would be additional to the road and rail system already existing; but the latter terminate in Venice, whereas the rapid transit railway, running in tunnel beneath the lagoon to the Island on which Venice is built, would continue under the lagoon to the holiday resort of Jessolo, and so provide a through route between Jessolo, Venice, and towns on the mainland which include Padua, Treviso and Mestre.

UNITED KINGDOM

MANCHESTER: Following proposals put forward by consultants, Manchester Corporation, it is reported, is to seek powers to construct an eleven-mile rapid transit railway at a cost of £50.m The recommended route is between Higher Blackly to the north of the city, via the city centre to Northenden in the south, with a short easterly spur to East Didsbury which would continue to a car depot beyond. Of the 11 miles of route, 1·6 miles would be in driven tunnel, 3·3 miles in tunnel of cut and cover construction, 2·0 miles on track elevated on single supporting pillars, and 3·0 miles on embankment. There would be fifteen stations, several close together on the in-town section, which would necessitate the use of cars with high acceleration characteristics.

LIVERPOOL: It was announced on March 5, 1970, that plans were to be made for a new underground railway for Liverpool. It would take the form of a two-mile single-tunnel loop, extending the Mersey tunnel line under the city centre and providing an underground link between busy stations on the British Railways surface system.

ISRAEL

HAIFA in Israel, which already has a mile-long funicular railway in rock tunnel running up the lower slopes of Mount Carmel, is said to be considering a rapid transit scheme, and so also is TEL AVIV, where at present urban and inter urban bus services carry 90 per cent. of the city's incoming and outgoing passengers.

The author concludes the main text of this book with references to some cities with railways which, although running underground for short distances and possessing underground stations serving central areas, cannot be described as integral rapid transit railways. Those mentioned here are, with the exception of the railway serving Athens, parts of the country's main line railway systems.

The Athens tunnel line is part of the 12 km Hellenic Electric Railways line which runs between the port of Piraeus and Athens. It descends from the surface into a 2·4 km long, shallow tunnel under the city and serves three underground stations, the most important of which is Omania Square.

In Brussels an underground section of railway known as the Brussels Junction Railway was built to connect the Belgian State Railway's network north of the city with that to the south. A two-kilometre tunnel of considerable width carries heavy internal and some international rail traffic over three pairs of tracks between the Nord and Midi main line stations. Along the tunnel there are three intermediate stations.

In Copenhagen the suburban lines of the Danish State Railways are carried through a tunnel section 1·6 km long which links the city's northern rail network with that to the south. The tunnel carries four tracks, two for diesel and two for electrified services.

Naples has an underground electrified railway line which was built partly as an alternative approach to Naples from Rome whilst the Rome–Naples direct rail route was being built. The line was also intended to be the nucleus of an underground railway serving Naples, but the project was not developed. The tunnel carries both main line and local stopping trains. On the actual tunnel section, which lies between Naples Central and Mergellina stations and is 5·5 km long, there are three underground stations.

In Liverpool, UK, a tunnel section of the British Railways system carrying electrified rail traffic under the River Mersey provides a rail link between Liverpool on one side and Birkenhead on the other. The tunnel line divides at Hamilton Square station on the west bank to continue north westward to Park and southward to Rock Ferry. Rail traffic includes through trains from Liverpool to New Brighton and West Kirby via Birkenhead Park. The total length of route actually in tunnel (i.e. that between Liverpool Central and Birkenhead Central, and Park) is 3¼ miles (5·2 km).

In Warsaw an underground section of railway carries the electrified suburban rail services of the Polish State Railways beneath the city in an east–west direction. The tunnel, lying between Warsaw Srodmiescie station and the River Vistula, is 4·5 km long and carries four tracks. Along it there are three underground stations.

SHORT GLOSSARY OF
TECHNICAL TERMS

Automatic block signalling: Incorporates safety devices, based on track circuiting, which ensures that a signal behind a train is automatically put to danger. Essentially, running rails are divided into sections of track electrically insulated from each other. Signal current passes along one rail, through a signal relay and back through the other rail. Current passing through the relay closes its contacts for a 'proceed' signal. A train entering a track circuited section provides a shorter path for the signal current through its wheels and axles. Consequently the relay is de-energized, causing the signal to go to danger, where it remains until the train passes beyond a safe 'overlap' (clearance point), after which the relay is re-energized and the proceed signal restored.

Automatic couplers: A coupler fitted to rolling stock that locks the coupling normally by means of a tongue and slot, and at the same time couples the cars' electrical, mechanical and pneumatic connections.

Catenary system: Refers in railway terms to the complex of overhead wires carrying current, or suspending those wires.

Cut and Cover: A method of trenching shallow tunnel where the side walls are sometimes first dug as deep trenches and filled with a clay slurry to retain the trench walls. The slurry is displaced (to be used again) by poured concrete around reinforcing rods. Sometimes the roof is similarly cast and the street surface relaid temporarily, allowing the core of earth to be removed subsequently.

Dead Man's Handle: A safety device incorporated in the driver's controller to ensure against the possible consequence of illness or collapse. Should the driver lift his hand from the controller handle, traction current is cut off and brakes are applied.

Dynamic braking: A method of electrical braking where the driving

180

motors are used as generators, thus exercising a retarding force. in some forms the electrical energy generated is dissipated through rheostats (adjustable electrical resistances) and in others, usually referred to as regenerative it is returned to the supply system.

Electronic ticket apparatus: The equipment in London includes ticket encoding and ticket reading apparatus. Tickets with magnetic oxide backing have journey details encoded on them which are 'read' by electronic equipment in the barrier gates. A valid ticket placed in a slot causes a first pair of doors to open, and as a passenger steps through a second pair of doors open and the first pair close behind him. A continuous stream of passengers inserting valid tickets one after the other causes the doors to remain open, giving an unimpeded flow. Inwards, the ticket is returned to the passenger. At outward barriers the ticket is retained. There are variations of basically-principled equipment in operation elsewhere.

Grade: The American equivalent used when referring to railway level (grade) crossings.

Multiple-unit Rolling Stock: This developed from early trains consisting of electric locomotives and trailer coaches. Trains with motors installed only in the end cars were followed by trains with intermediate cars motored, and lastly by all-motored cars, with relatively increased speed and acceleration characteristics. The motors throughout the train are under a single control.

Pantograph: Often used to describe the spring-loaded bracket on car roofs that keeps the car's electrical supply collecting equipment in contact with the overhead wire system.

Power Station Capacity: This is generally referred to in kilowatts, electrical units of power each equalling 1·34 horse-power. The recently modernized London Transport power station at Lots Road, equipped with oil-fired boilers and turbo-alternators, has a capacity of up to 180,000 kilowatts, sufficient to supply a town the size of Eastbourne.

Programme Machines: This apparatus consists basically of a plastic roll which moves in steps, each corresponding to one train movement. In the roll are punched holes which constitute a code of information concerning the details of routes and destinations of trains in timetable sequence. Probed contacts pressed against the roll read particulars of a train approaching a junction and accordingly signal it. The train, proceeding, energizes a relay and motivates apparatus which lifts the contacts clear of the coded roll whilst a motor rotating the roll moves it one step. The contacts are once again pressed against the roll, after the train movement, ready for the next train movement. In practice, the description of trains dispatched from a particular point are received and stored by apparatus. The programme machine checks and releases in timetable frequency information that is fed electrically to interlocking machines, which automatically operate points or switches. Other apparatus based on timing, under the supervision of regulating staff, is linked with the sequence programme machines so that such contingencies as train delays or cancellations can be dealt with. The foregoing relates to equipment in use on the London system, but in New York and elsewhere similar apparatus is in use.

Railway ties: American equivalent to railway sleepers.

Shield for driven tunnel: Several types of shield have been used for running tunnel excavation, including the drum-digger type, which basically consists of an outer non-revolving cylinder or drum, with a cutting edge which is rammed forward, and an inner revolving drum which carries cutting teeth. Another type has a cylindrical body like the former type, but inside there is a rotating cutting wheel with four spokes. Greathead-type, large-diameter, non-revolving shields are used for station tunnel excavation, the shield in all three types being thrust forward by rams or jacks bearing against completed tunnel linings.

Speed Control Signalling: The apparatus employed for automatically controlling the speed of manually-driven trains employs an electromagnetic detector which instantly measures a train's speed and actuates a relay, which in turn controls a line-side signal. It is employed in London to permit close operation of trains consistent with absolute safety, but is also installed generally where conditions call for speed regulation.

Schnellbahn: Fast or rapid railway.

S Bahn: Stadt or Town railway.

Step-plate junction: A junction of tunnel lines where the rings forming the tunnel widen or are reduced in size (in steps) to enlarge or reduce the tunnel diameter.

Sub-Stations: In London the old rotary converter units in sub-stations are being replaced by modern static rectifier and trans-former units. The former's function is to convert alternating current into direct current and the latter's to transform the electrical pressure or voltage; for example, from high voltage to lower voltage for motive power. Modernization of power distri-bution is taking place on many of the long-established systems, where older manned sub-stations are being replaced by un-manned, automatically operated sub-stations.

Theodolite: An instrument used by surveyors for measuring horizontal and vertical angles.

Train-Stop: A 'train-stop' located at the track-side has an arm which is down when the signal ahead is clear. The arm rises when the signal goes to danger. If a train attempts to pass the danger signal the raised arm engages a trip cock on the train, which releases compressed air from the 'train-pipe', causing brakes to be applied. The 'stop' has built-in proving circuits to guard against possible failure of the apparatus.

Truck: American equivalent to the English railway bogie.

Tunnel Alignment: Construction of new tube tunnel starts with sinking vertical shafts at intervals along the route. From tunnel headings dug at right-angles to the future alignment a surface sight line, exactly over the line of route, is taken by theodolite and transferred to tunnel level by hanging plumb lines on oppo-site sides of the vertical shaft, aligning them with the surface line at top and bottom, and measuring the correct distance along the heading and the correct depth to find the centre line of the running tunnel.

Tunnel invert: The reverse of a tunnel arch which in a running tunnel contains the track bed.

V Bahn: Verbunden Bahn or connecting railway.